How to Becom
Surveyor

Thinking about a career in property or construction? Thinking of becoming a Chartered Surveyor? *How to Become a Chartered Surveyor* demystifies the process and provides a clear road map for candidates to follow.

The book outlines potential pathways and practice areas within the profession and includes the breadth and depth of surveying, from commercial, residential and project management, to geomatics and quantity surveying. Experienced APC assessor and trainer, Jen Lemen BSc (Hons) FRICS, provides invaluable guidance, covering:

- routes to becoming a Chartered Surveyor, including t-levels, apprenticeships and alternative APC routes such as the Senior Professional, Academic and Specialist assessments
- areas of professional practice
- advice for the AssocRICS, APC (MRICS), FRICS and Registered Valuer assessments, including both written and interview elements
- advice on referrals and appeals
- how to support candidates, including the role of the Counsellor and Supervisor
- opportunities for further career progression, including further qualifications and setting up in practice as an RICS regulated firm
- global perspectives
- professional ethics for surveyors

Written in clear, concise and simple terms and providing practical advice throughout, this book will help candidates to decode and understand the RICS guidance, plan their career and be successful in their journey to become a Chartered Surveyor. It will also be of relevance to academic institutions, employers, school leavers, apprentices, senior professionals, APC Counsellors/Supervisors and careers advisors.

Jen Lemen BSc (Hons) FRICS is a Chartered Surveyor and co-founder of Property Elite, a company providing training and support to RICS APC, AssocRICS and FRICS candidates. She is also an experienced property consultant (co-Director of Projekt), an RICS Registered Valuer and an RICS Accredited Mediator, with nearly 15 years' experience working in the commercial property sector. She also has academic experience as a Senior Lecturer at the University of the West of England, Lecturer at the University of Portsmouth and Associate Tutor at the University College of Estate Management. Her RICS assessment experience includes sitting on final APC interview panels, APC appeal panels and being a lead APC preliminary review assessor. She also assesses written AssocRICS submissions.

How to Become a Chartered Surveyor

Jen Lemen BSc (Hons) FRICS

LONDON AND NEW YORK

First published 2022
by Routledge
2 Park Square, Milton Park, Abingdon, Oxon OX14 4RN

and by Routledge
605 Third Avenue, New York, NY 10158

Routledge is an imprint of the Taylor & Francis Group, an informa business

British Library Cataloguing-in-Publication Data
A catalogue record for this book is available from the British Library

Library of Congress Cataloging-in-Publication Data
Names: Lemen, Jen, author.
Title: How to become a chartered surveyor / Jen Lemen.
Description: Abingdon, Oxon ; New York, NY : Routledge, 2021. | Includes bibliographical references and index.
Identifiers: LCCN 2021003901 (print) | LCCN 2021003902 (ebook) | ISBN 9780367742270 (hbk) | ISBN 9780367742195 (pbk) | ISBN 9781003156673 (ebk)
Subjects: LCSH: Surveying--Vocational guidance.
Classification: LCC TA549 .L455 2021 (print) | LCC TA549 (ebook) | DDC 526.9023--dc23
LC record available at https://lccn.loc.gov/2021003901
LC ebook record available at https://lccn.loc.gov/2021003902

ISBN: 978-0-367-74227-0 (hbk)
ISBN: 978-0-367-74219-5 (pbk)
ISBN: 978-1-003-15667-3 (ebk)

DOI: 10.1201/9781003156673

Typeset in Baskerville
by Taylor & Francis Books

This book is dedicated to those who have been a part of my journey to becoming a Chartered Surveyor. My family; Mum and Dad, Granny and Grandad Birmingham, Granny and Grandad Scotland. My business partner and mentor through life; Rachel Saint. My wife and soulmate, Frankie, and my new Bournemouth family. And not forgetting my two best canine friends, Ted and Ben. I wouldn't be here without you all.

Contents

Illustrations

Figures

Tables

List of abbreviations

APC	Assessment of Professional Competence
ARC	Assessment Resource Centre
AssocRICS	Associate Member of the Royal Institution of Chartered Surveyors
BIM	Building Information Modelling
BPS	Basic Payments Scheme
CHP	Complaints Handling Procedure
CPD	Continued Professional Development
CRE	Corporate Real Estate
CVR	Cost Valuation Reconciliations
DBS	Desk Based Review
DRS	Dispute Resolution Service
GCSE	General Certificate of Secondary Education
EMEA	Europe, Middle East and Africa
FIDIC	International Federation of Consulting Engineers
GIS	Geographic Information Systems
FM	Facilities Management
FRICS	Fellow of the Royal Institution of Chartered Surveyors
FRA	Fire Risk Assessment
GDPR	General Data Protection Regulation
HMO	House in Multiple Occupation
ICSM	International Construction Measurement Standards
IFSS	International Fire Safety Standards
ILMS	International Land Measurement Standards
IPMS	International Property Measurement Standards
IVS	International Valuation Standards

JCT	Joint Contracts Tribunal
MEES	Minimum Energy Efficiency Standard
MENA	Middle East and North Africa
MRICS	Member of the Royal Institution of Chartered Surveyors
NEC	New Engineering Contract
NRM	New Rules of Measurement
PGDip	Post Graduate Diploma
PhD	Doctor of Philosophy
PII	Professional Indemnity Insurance
PS	Professional Standards
RIBA	Royal Institute of British Architects
RICS	Royal Institution of Chartered Surveyors
RRV	Regulatory Review Visit
SCSI	Society of Chartered Surveyors Ireland
UCAS	Universities and Colleges Admissions Service
UN	United Nations
VPGA	Valuation Practice Guidance Applications
VPS	Valuation Technical and Performance Standards
VRS	Valuer Registration Scheme

Foreword

If you have picked up this book, then it's likely that you have an interest in becoming a Chartered Surveyor. If you don't already, we hope that you will be inspired to, by reading to the end.

The author of this book never intended to become a Chartered Surveyor. After leaving school she initially studied Sports & Exercise Science full-time at the University of Bath. However, this choice became questionable to her due to the full-time mode of study and the lack of perceived career opportunities after graduating. After a year and a half, the author decided to leave the course and reconsider her options.

The author's father and grandfather were both Chartered Surveyors. George Lemen FSVA, left school at the age of 14 and joined the Property Department of Fleming Reid & Co. as an office junior. After World War II ended, he re-joined the company and qualified by correspondence course with the Incorporated Society of Auctioneers and Landed Property Agents (since incorporated into RICS). He eventually joined H. Samuel Jewellers where he spent the majority of a very successful career as their Estates Director.

Like father, like son. The author's father, Don Lemen FRICS, left school and then joined Edwards Bigwood & Bewlay as a trainee surveyor. He qualified as a Chartered Surveyor by the College of Estate Management correspondence course. He recently retired having spent the majority of his career working with Land Securities, Donaldsons, DTZ and Cushman & Wakefield, specialising in shopping centre property and facilities management.

Having positive surveying role models inspired the author to consider a career as a Chartered Surveyor. Growing up, she had also enjoyed the opportunity and experience of spending time with her father when he was working, visiting shopping centres both in the UK and abroad. The next step, of course, was to find a way to avoid full-time study, as the author wanted to earn a salary whilst gaining practical experience.

The University of the West of England, at the time, offered a five year part-time course in Property Management & Investment. Alongside joining the course, the author found employment as a trainee surveyor at Seagrove & Lambert Surveyors in Bristol. She pursued the course on a day release basis, before accepting a new role at Rapleys, also based in Bristol. After two more years, she converted her mode of study to distance learning without attendance, allowing her to take on more responsibilities and gain wider work experience.

The author graduated with a First Class degree and an overall average of 91%. She was also awarded the RICS South West Regional Prize for high achievement.

A year later, the author was proud to qualify as a Chartered Surveyor through the Royal Institution of Chartered Surveyors (RICS) Assessment of Professional Competence (APC). This process was not easy, with the intense combination of work and studying before preparing for a final professional interview.

After qualifying as a Chartered Surveyor, the author proceeded through the career ranks before leaving to set up in practice with her business partner and career-long mentor, Rachel Saint DipSurv MRICS. Alongside setting up a RICS regulated surveying firm, Projekt, the pair founded a highly successful training and support company for prospective Chartered Surveyors, Property Elite.

The author also attained a higher level of surveying recognition, becoming a Fellow of the RICS (FRICS) and Accredited Mediator, whilst also lecturing and supervising dissertation students at various universities.

Safe to say, a career in Chartered Surveying was an extremely sound decision for the author and she has never looked back. Continue reading to find out more about how you can become a Chartered Surveyor too.

1 What Is a Chartered Surveyor?

Introduction

Chartered Surveying is a diverse profession, including a wider breadth and depth of roles and industries than perhaps even many Chartered Surveyors themselves appreciate.

The description, Chartered Surveyor, is a very special term. Whilst anyone can call themselves a surveyor, only very specific professionals can use the term Chartered Surveyor. This encompasses the Royal Institution of Chartered Surveyors (RICS) designations, Member (MRICS) and Fellow (FRICS). There is a third designation, AssocRICS, which is highly regarded but does not confer Chartered status.

In this chapter, we will look at what a Chartered Surveyor is and how this term came into use.

Chapter 2 will further investigate the areas of professional practice within which Chartered Surveyors operate.

The Role of the Chartered Surveyor

As a starting point, we will take a contemporary look at the responsibilities and career prospects of a Chartered Surveyor.

Chartered Surveyors are involved in a wide variety of roles across built and natural environments. This includes the measurement, valuation, protection and enhancement of global physical assets. Responsibility for this spans across the entire building lifecycle; from purchasing land, to the planning process and through to obsolescence and future redevelopment.

DOI: 10.1201/9781003156673-1

Surveyors are employed across a wide range of private firms, public sector bodies and charities. They can be found in consultancy or advisory roles, e.g., a building surveyor, quantity surveyor, project manager, agency surveyor or development surveyor. They may also be found in client-side roles, e.g., as a property director or estates surveyor.

Even more so in 2020 onwards, the role of a Chartered Surveyor is extremely varied with work being undertaken in the office, on site and at home. Careers may be global, national or local, with the opportunity to travel and work on projects of all sizes, shapes and types. It is fair to say that no two days are the same in the life of a Chartered Surveyor.

These are some of the benefits of pursuing a career in Chartered Surveying:

- Making a difference to the economy, environment and community. Chartered Surveyors create a legacy through their work, including being involved with iconic buildings, protecting the natural environment, and helping consumers to purchase new homes.
- Earning a good salary (see Career Prospects below for more details) with a variety of different ways to become qualified.
- Excellent career prospects with continual progression and opportunities to work in a variety of sectors and markets.
- Good work life balance, with the opportunity to work flexible hours in some organisations and companies – it is certainly not just a desk job.
- Opportunities to travel and work in different environments or countries.
- Sociable work with an emphasis on collaboration and teamwork to provide the highest standards of professional advice to clients.
- Requires a mix of soft and technical skills, including being personable and approachable.
- Varied roles and opportunities to suit a range of different personalities and skillsets.

Career Prospects

The career prospects for Chartered Surveyors vary widely, with salaries depending on location, industry, experience and specialisation.

Graduate salaries tend to range between £20,000 to £30,000 (RICS, 2020a). There is a clear step up in salary when surveyors become Chartered. The average difference has been reported as £13,600 (39% difference) (RICS and Macdonald & Company, 2019) and £16,000 (RICS, 2020a), demonstrating the financial benefit of becoming a Chartered Surveyor.

For Chartered Surveyors, the average median UK salary was £48,000 (RICS and Macdonald & Company, 2019). However, salaries are often commission based, particularly in agency or investment-related roles. This means that some surveyors will earn salaries of £100,000 plus. Typical package benefits in larger firms include private health insurance, life insurance, pension contributions and a company car or car allowance.

The median salary differs between genders; £50,000 for men and £41,685 for women (RICS and Macdonald & Company, 2019). This demonstrates that the industry continues to suffer from a gender pay gap, although this is being addressed by a variety of industry initiatives including the RICS Inclusive Employer Equality Mark and diversity and inclusion strategy.

Chartered Surveying in 2020

RICS statistics (RICS, 2020b) show that there are approximately 110,000 RICS qualified surveyors globally (AssocRICS, MRICS and FRICS).

Around 77,000 of these are in the UK; approximately 84% are male, 15% are female and under 1% did not identify as either, preferring not to state their gender.

The Royal Institution of Chartered Surveyors

The Royal Institution of Chartered Surveyors (RICS) is the global governing body for Chartered Surveyors. The RICS was originally formed as the Surveyors Club in 1792 and developed over time into a professional association representing surveyors and the property profession. This reflected the rise of industrialisation and expansion of infrastructure, housing and transport, which required advice from professionals and created a need for some form of regulation.

The Surveyors' Institution was founded on 15th June 1868 and later incorporated by Royal Charter on 26th August 1881. On 27th October 1930, the name was changed to The Chartered Surveyors'

Institution, which later became what we now know as the RICS on 3rd July 1947. The motto of the RICS is 'Est modus in rebus' or, 'There is measure in all things'.

Royal Charter

The Royal Charter sets out the fundamental difference between a surveyor and a Chartered Surveyor. It incorporates the Original Royal Charter (dated 26th August 1881) and Supplemental Royal Charter (which amends the Original Royal Charter, with the latest changes being made in February 2020).

The Royal Charter means that changes to the RICS constitution, known as the Bye-Laws, have to be approved in a two-step process. Firstly, they have to be approved by a majority members' vote at a general meeting. Secondly, they have to be ratified by the Privy Council, which is part of the UK Government.

The Royal Charter states that 'Chartered Members', i.e., Chartered Surveyors, may only be Fellows (FRICS) and Professional Members (MRICS) of the RICS. It also gives certain firms the right to use the designation, Chartered Surveyors. Chartered Surveyors and firms of Chartered Surveyors are required by the Royal Charter to follow the RICS Bye-Laws and Regulations.

The Royal Charter states that the aim of the RICS is 'to maintain and promote the usefulness of the profession for the public advantage in the United Kingdom and in any other part of the world' (RICS, 2020c). In particular, this focusses on protecting the public and upholding the reputation of the RICS within society.

The Royal Charter sets a gold standard of excellence and integrity for the industry, particularly because Royal Charters remain in demand from a wide variety of professions. Other professionals with a Royal Charter in the UK include Chartered Accountants, Chartered Managers, Chartered Psychologists, Chartered Structural Engineers and Chartered Town Planners.

Self-Regulation

The fact that the RICS has a Royal Charter means that the Chartered Surveying profession operates under a model of self-regulation. This means that there is no Government regulation of Chartered Surveyors.

Instead, they are internally regulated by the RICS through the Bye-Laws and RICS Regulation.

The Royal Charter means that the Government should be confident that Chartered Surveyors are regulated appropriately and diligently, in line with the principles of better regulation; Proportionality, Accountability, Consistency, Targeting and Transparency. These are set out by the UK Cabinet Office's Better Regulation Commission and adopted by the RICS in its regulatory model.

The ability to self-regulate is beneficial because it avoids the time and cost of Government introducing and maintaining appropriate legislation. Effectively, there is no need to legislate because the Government is confident that through the Royal Charter, the RICS is internally regulating at arm's length, as a business and in line with modern working practices.

Bye-Laws and Regulations

The RICS Bye-Laws (RICS, 2020d) and RICS Regulations (RICS, 2020e) were both last updated in February 2020. Whilst the Bye-Laws set out more general regulatory principles and procedures, the Regulations set out specific details relating to the operation of the RICS as a regulatory body.

Together, these set out the regulatory requirements of the RICS, including issues such as membership eligibility, registration of firms, use of designations, subscriptions and fees, conduct, powers and governance structure. The way that the RICS regulates is further guided by written Governance Procedures and Processes (also known as Standing Orders) (RICS, 2020f).

The key takeaway from the Bye-Laws and Regulations, is that Chartered Surveyors must conduct themselves in a 'manner befitting membership of the RICS'. Throughout this book, we will look at how Chartered Surveyors can meet this requirement in terms of acting ethically, responsibly and professionally.

Conclusion

In conclusion, Chartered Surveying is a highly regulated industry with clear top-level guidance and stringent requirements set out by the RICS. The profession has been around for a long time and there is continued evidence of demand for Chartered Surveyors across both the built and natural environment sectors.

In Chapter 2, we will explore the width and diversity of roles within which Chartered Surveyors operate, before looking more closely at how you can become a Chartered Surveyor.

2 Areas of Professional Practice

Introduction

We have already taken an in-depth look at the history and current state of play in relation to the surveying industry.

In this chapter, we take a closer look at the areas of professional practice that Chartered Surveyors may advise or work within.

What Types of Chartered Surveyor Are There?

Chartered Surveyors work across the lifecycle of built and natural environments. Frequently, they work alongside and collaborate with other professionals, such as architects, structural engineers, bankers, ecologists, town planners, property developers and senior managers, e.g. Chief Financial Officers or Managing Directors. Some surveyors may also be dual or multi qualified, for example they may be a Chartered Surveyor and a Chartered Town Planner.

Surveying roles are traditionally split into three main sectors; construction and infrastructure, property, and land.

What Does the Construction and Infrastructure Sector Include?

The Construction and Infrastructure Sector typically relates to the physical construction and building of infrastructure (e.g., roads, bridges, railways, energy supplies and telecoms) and buildings.

Buildings can be minor or major projects, from a single residential dwelling to major projects such as the Olympic Park or HS2. Typical

DOI: 10.1201/9781003156673-2

roles in the Construction and Infrastructure Sector include Building Surveyors, Project Managers, Quantity Surveyors, Building Control Surveyors and Infrastructure Surveyors.

What Does the Property Sector Include?

The Property Sector includes both residential (domestic) and commercial (non-domestic) buildings, as well as trading businesses and personal property, e.g., fine art and antiques.

Typical roles in this sector include Property Surveyors, Valuation Surveyors, Management Consultancy Surveyors and Facilities Management Surveyors.

What Does the Land Sector Include?

The Land Sector relates to town and environmental planning, climate change and resource use.

Typical roles in this sector include Geomatics Surveyors, Environmental Surveyors, Minerals and Waste Surveyors, Rural Surveyors and Planning and Development Surveyors.

How Are the Different Types of Chartered Surveyor Defined by RICS?

The Royal Charter (RICS, 2020c), states that the RICS aims to 'secure the advancement and facilitate the acquisition of that knowledge which constitutes the profession of a surveyor, namely, the arts, sciences and practice of:

a determining the value of all descriptions of landed and house property and of the various interests therein and advising on direct and indirect investment therein;
b managing and developing estates and other business concerned with the management of landed property;
c securing the optimal use of land and its associated resources to meet social and economic needs;
d surveying the fabric of buildings and their services and advising on their condition, maintenance, alteration, improvement and design;
e measuring and delineating the physical features of the Earth;

- f managing, developing and surveying mineral property;
- g determining the economic use of resources of the construction industry, and the financial appraisal, management and measurement of construction work;
- h selling (whether by auction or otherwise) buying or letting, as an agent, real or personal property or any interest therein.'

There are nineteen specific designations or titles that Chartered Surveyors can use, apart from the simple title, Chartered Surveyor, which are defined by the RICS Regulations (RICS, 2020e):

1. Chartered Arts and Antiques Surveyor;
2. Chartered Building Surveyor;
3. Chartered Building Control Surveyor;
4. Chartered Commercial Property Surveyor;
5. Chartered Construction Surveyor;
6. Chartered Engineering Surveyor;
7. Chartered Environmental Surveyor;
8. Chartered Facilities Management Surveyor;
9. Chartered Forestry Surveyor;
10. Chartered Hydrographic Surveyor;
11. Chartered Land Surveyor;
12. Chartered Machinery Valuation Surveyor;
13. Chartered Management Consultancy Surveyor;
14. Chartered Minerals Surveyor;
15. Chartered Planning and Development Surveyor;
16. Chartered Project Management Surveyor;
17. Chartered Quantity Surveyor;
18. Chartered Valuation Surveyor;
19. Chartered Civil Engineering Surveyor.

The RICS, as an organisation and governing body, is further structured to include seventeen professional groups. These relate to the technical specialisms of different types of Chartered Surveyor and include:

1. Building Control;
2. Building Surveying;
3. Commercial Property;

4. Dispute Resolution;
5. Environment and Resources;
6. Facilities Management;
7. Geomatics;
8. Machinery and Business Assets;
9. Management Consultancy;
10. Minerals and Waste Management;
11. Planning and Development;
12. Personal Property/Arts and Antiques;
13. Project Management;
14. Quantity Surveying and Construction;
15. Residential Property;
16. Rural;
17. Valuation.

The RICS also operate five specialist sector forums which are responsible for producing technical guidance for other Chartered Surveyors:

1. Building Conservation;
2. Dilapidations;
3. Infrastructure;
4. Insurance;
5. Telecoms.

From the above, it is clear that surveying is a diverse profession with many areas of potential specialism. Chartered Surveyors are often able to work within many different areas of practice during their careers, sometimes starting off in wide 'general practice' roles and later, choosing to specialise. Others may start their careers in a niche role and they will quickly become specialists in that given area.

Chartered Surveyors can generally, therefore, find a role and specialism which interests and motivates them rather than becoming siloed within one specific area of practice. This, therefore, allows for personal and professional growth and development; resulting in an ongoing rewarding and engaging career.

How Do these Various Sectors, Specialisms and Designations Relate to Becoming Qualified as a Chartered Surveyor?

We will discuss the routes, which are generally based on qualifications, experience and seniority, to becoming a Chartered Surveyor in the later chapters. However, given the wide breadth and depth of the surveying profession, these routes are split into a number of sector-specific pathways, e.g., Building Surveying and Valuation.

To an extent, these pathways define the type of Chartered Surveyor that you may become by setting a sector-specific framework of technical competencies that you must meet during your assessment to become a Chartered Surveyor.

The career of a Chartered Surveyor will be subject to dynamic change over time. For example, just because a Chartered Surveyor qualified on the Building Surveying pathway, it does not mean that they cannot work in another sector or on different types of instruction. This would instead depend on if they are sufficiently skilled and experienced to do so and have appropriate Professional Indemnity Insurance cover. This means that someone who qualified as a Building Surveyor could eventually become employed as a Quantity Surveyor, or move into property development, for example.

There are twenty-two sector pathways for the RICS Assessment of Professional Competence (APC), which is the written and interview assessment that leads to MRICS status, i.e., to become a Chartered Surveyor. Candidates pursuing the APC must choose a sector pathway which most closely reflects their current experience, role and knowledge. This is because each pathway includes a specific range of technical competencies and candidates will need to demonstrate their practical experience.

There is another qualification, AssocRICS, which does not lead to Chartered Surveyor status, but provides an entry-level professional qualification which can later be progressed to MRICS via a separate assessment. For the AssocRICS qualification, there are fourteen sector pathways available for candidates to choose from, fewer in comparison to the APC.

We will look at each of these sector pathways in turn below, along with some of the technical competencies that candidates will need to have experience of:

1. Building Control;
2. Building Surveying;
3. Facilities Management;
4. Geomatics;
5. Infrastructure;
6. Land and Resources;
7. Project Management;
8. Quantity Surveying and Construction;
9. Valuation;
10. Commercial Real Estate (AssocRICS has two separate pathway options);
11. Residential (AssocRICS has three separate pathway options);
12. Corporate Real Estate (MRICS only);
13. Environmental Surveying (MRICS only);
14. Management Consultancy (MRICS only);
15. Minerals and Waste Management (MRICS only);
16. Personal Property/Arts and Antiques (MRICS only);
17. Research (MRICS only);
18. Rural (MRICS only);
19. Taxation Allowances (MRICS only);
20. Valuation of Businesses and Intangible Assets (MRICS only).

Candidates should refer back to the relevant pathway guide when considering how the core and optional technical competencies fit with their role and experience. This is because similarly titled competencies may have different descriptions, knowledge or required experience based on the specific pathway. There are also some competencies with titles that may need further investigation or reading, e.g. Big Data or Business Case, in order for candidates to understand what is required of them.

Building Control

Building Control Surveyors advise clients on the building regulations and other design and construction-related legislation. This could relate to either new buildings, or buildings requiring redevelopment, refurbishment, or to be demolished or extended. They may give advice on meeting minimum standards, or where unexpected issues are encountered on site.

Core technical competencies include:

1 Building Control Inspections;
2 Fire Safety;
3 Inspection;
4 Legal/Regulatory Compliance.

Other areas of experience include technical competencies such as:

1 Building Information Modelling (BIM) Management;
2 Building Pathology;
3 Conservation and Restoration;
4 Construction Technology and Environmental Services;
5 Contaminated Land;
6 Measurement;
7 Planning and Development Management;
8 Risk management;
9 Sustainability;
10 Works Progress and Quality Management.

Building Surveying

Building Surveying encompasses a wide range of advice on property and construction, including specialist areas such as party walls, rights of light, dilapidations and conservation.

Core technical competencies include:

1 Building Pathology;
2 Construction Technology and Environmental Services;
3 Contract Administration;
4 Design Specification;
5 Inspection;
6 Legal/Regulatory Compliance;
7 Fire Safety.

Other areas of experience include technical competencies such as:

1 BIM Management;
2 Commercial Management;

3 Conservation and Restoration;
4 Contract Practice;
5 Design Economics and Cost Planning;
6 Development/Project Briefs;
7 Housing Maintenance, Repairs and Improvements;
8 Maintenance Management;
9 Insurance;
10 Landlord and Tenant;
11 Measurement;
12 Procurement and Tendering;
13 Project Finance, Quantification and Costing;
14 Risk Management;
15 Works Progress and Quality Management.

Commercial Real Estate

Commercial Real Estate is typically split into a number of sub-sectors; retail, offices, industrial and leisure. However, there are many emerging alternative sectors and sub-sectors, including flexible, serviced offices, as well as logistics and distribution centres.

Chartered Surveyors in this sector are sometimes known as General Practice Surveyors, i.e., they advise on a wide range of instruction types and sectors. Others focus specifically on areas of advice, such as for agencies (purchase, sale, leasing and letting), property management, landlord and tenant (also known as lease advisory or lease consultancy, or even professional services in more traditional firms), telecoms, valuation, investment and development.

Core technical competencies include:

1 Inspection;
2 Measurement;
3 Valuation.

Other areas of experience include technical competencies such as:

1 Auctioneering;
2 BIM Management;
3 Building Pathology;
4 Capital Taxation;

5 Compulsory Purchase and Compensation;
6 Contaminated Land;
7 Corporate Recovery and Insolvency;
8 Development Appraisals;
9 Insurance;
10 Indirect Investment Vehicles;
11 Investment Management;
12 Landlord and Tenant;
13 Leasing and Letting;
14 Legal/Regulatory Compliance;
15 Loan Security Valuation;
16 Local Taxation/Assessment;
17 Planning and Development Management;
18 Property Finance and Funding;
19 Property Management;
20 Purchase and Sale;
21 Strategic Real Estate Consultancy.

On this pathway, 30% of a candidate's experience can come from other property sectors, e.g., residential or rural. This reflects the fact that many properties and portfolios are of mixed use, i.e., not solely comprising commercial buildings.

Corporate Real Estate

Corporate Real Estate (CRE) sounds very similar to Commercial Real Estate; however, the two pathways are very different. Both involve advising on commercial property matters though.

CRE relates to advising on the entire lifecycle of a property portfolio, typically owned by a private or public sector organisation. This can include advising on strategic planning, portfolio analysis, property requirements and management of a portfolio. The aim being to optimise occupational costs and maximise the utilisation of available space. This typically requires an in-depth understanding of wider business objectives and operational requirements to ensure alignment with the property portfolio.

There is also cross over between CRE and Facilities Management, particularly in relation to advising on operational costs and use of space.

Core technical competencies include:

1 Business Alignment;
2 Strategic Real Estate Consultancy;
3 Business Case;
4 Landlord and Tenant;
5 Property Management;
6 Valuation.

Other areas of experience include technical competencies such as:

1 Change Management;
2 Inspection;
3 Leasing and Letting;
4 Local Taxation/Assessment;
5 Measurement;
6 Performance Management;
7 Procurement and Tendering;
8 Programming and Planning;
9 Purchase and Sale;
10 Supplier Management;
11 Workspace Strategy.

Environmental Surveying

Environmental Surveyors are involved in environmental management, monitoring and assessment. This could include involvement with contaminated land, environment audits, due diligence, insurance, pollution control, renewable resources, forestry, woodland management, Geographic Information Systems (GIS), waste management and risk management.

Core technical competencies include:

1 Environmental Management;
2 Legal/Regulatory compliance.

Other areas of experience include technical competencies such as:

1 Consultancy Services;

2 Contaminated Land;
3 Environmental Assessment;
4 Environmental Audit and Monitoring;
5 Environmental Science and Processes;
6 Inspection;
7 Management of the Natural Environment and Landscape;
8 Planning and Development Management.

Facilities Management

Facilities Managers are responsible for managing the services and infrastructure that support organisations' core businesses. This requires the ability to manage change and mobilise resources, particularly through the use of innovative and digital technology.

Core technical competencies include:

1 Asset Management;
2 Business Alignment;
3 Client Care;
4 Legal/Regulatory Compliance;
5 Maintenance Management;
6 Performance Management;
7 Procurement and Tendering;
8 Project Finance;
9 Supplier Management;
10 Workspace Strategy.

Other areas of experience include technical competencies such as:

1 Big Data;
2 BIM Management;
3 Business Case;
4 Change Management;
5 Commercial Management;
6 Construction Technology and Environmental Services;
7 Consultancy Services;
8 Contract Administration;
9 Contract Practice;
10 Design and Specification;

11 Environmental Management;
12 Landlord and Tenant
13 Managing Projects;
14 Measurement;
15 Risk Management;
16 Smart Cities and Intelligent Buildings;
17 Stakeholder Management;
18 Strategic Real Estate Consultancy;
19 Waste Management;
20 Works Progress and Quality Management.

Geomatics

Chartered Surveyors specialising in Geomatics tend to use technology to collect, interpret, analyse and present spatial data relating to the natural and built environments. A strong understanding of land law and administration is required, together with the application of technical theory and digital tools.

Chartered Surveyors can achieve the designations Chartered Engineering Surveyor, Chartered Hydrographic Surveyor and Chartered Land Surveyor, providing they select the relevant technical competencies to the required level.

Core technical competencies include:

1. Cadastre and Land Administration;
2. Conflict Avoidance, Management and Dispute Resolution Procedures;
3. Engineering Surveying;
4. Geodesy, GIS, and Hydrographic Surveying;
5. Legal/Regulatory Compliance;
6. Measurement;
7. Remote Sensing and Photogrammetry;
8. Surveying and Mapping;
9. Surveying of Land and Sea;
10. Use of the Marine Environment.

Other areas of experience include technical competencies such as:

1. Big Data;

2 BIM Management;
3 Construction Technology and Environmental Services;
4 Consultancy Services;
5 Data Management;
6 Development/Project Briefs;
7 Environmental Assessment;
8 Ground Engineering and Subsidence;
9 Management of the Natural Environment and Landscape;
10 Planning and Development Management.

Infrastructure

Chartered Surveyors can advise on Infrastructure Projects across sectors such as energy, oil, gas, mining and water. This can include cost management and project management across the entire project life cycle.

Core technical competencies include:

1 Engineering Science and Technology;
2 Client Care;
3 Contract Practice;
4 Cost Prediction and Analysis;
5 Procurement and Tendering;
6 Programming and Planning;
7 Project Controls;
8 Quantification, Costing and Price Analysis;
9 Risk management.

Other areas of experience include technical competencies such as:

1 Asset Management;
2 BIM Management;
3 Compulsory Purchase and Compensation;
4 Contract Administration;
5 Cross Cultural Awareness in a Global Business Settting;
6 Leading Projects, People and Teams;
7 Managing Projects;
8 Project Finance and Stakeholder Management;
9 Supplier Management.

Land and Resources

The Land and Resources pathway has a focus on sustainability, balancing the needs of the entire built and natural environment lifecycle in terms of economic, environmental and social factors. There is an emphasis on the use of digital technology and effective conflict or dispute resolution.

The core technical competencies are very wide ranging and include:

1 Agriculture;
2 Big Data;
3 Cadastre and Land Administration;
4 Compulsory Purchase Compensation;
5 Geodesy and Minerals Management.

Management Consultancy

It might be surprising to find that Chartered Surveyors can also be Management Consultants, who typically advise on optimising business performance in a wide variety of organisations. Chartered Surveyors will often advise on maximising real estate strategy in order to meet wider corporate objectives. This requires strong skills and knowledge of economics, business and management, deployed through various consultancy services and approaches to meet clients' needs.

Core technical competencies include:

1 Business Case;
2 Business Planning;
3 Consultancy Services;
4 Research Methodologies and Techniques.

Other areas of experience include technical competencies such as:

1 Business Alignment and Change Management;
2 Corporate Finance;
3 Corporate Recovery and Insolvency;
4 Data Management;
5 Development Appraisals;
6 Development Projects/Briefs;
7 Economic Development;

8. Managing Resources;
9. Performance Management;
10. Programming and Planning;
11. Property Finance and Funding;
12. Smart Cities and Intelligent Buildings;
13. Strategic Real Estate Consultancy;
14. Workspace Strategy.

Minerals and Waste Management

Chartered Surveyors specialising in Minerals and Waste Management will advise from identification of an initial site opportunity to the future use of a former site. This requires specific legislative knowledge and consultative skills to engage with key stakeholders, e.g., landowners and the local authority.

Core technical competencies include:

1. Minerals Management;
2. Waste Management;
3. Legal/Regulatory Compliance;
4. Environmental Assessment;
5. Environmental Audit and Monitoring;
6. Ground Engineering and Subsidence;
7. Inspection;
8. Landlord and Tenant;
9. Legal/Regulatory Compliance;
10. Local Taxation/Assessment;
11. Planning and Development Management;
12. Surveying and Mapping;
13. Valuation.

Other areas of experience include technical competencies such as:

1. Consultancy Services;
2. Contaminated Land;
3. Contract Administration;
4. Development Appraisals;
5. Development Project Briefs;
6. Energy and Renewable Resources;

7 Risk Management;
8 Sustainability;
9 Works Progress and Quality Management.

Personal Property/Arts and Antiques

It may also be a surprise to learn that Chartered Surveyors can be involved in advising on Personal Property, such as fine art and antiques. This can involve valuations, disposals (i.e., selling), acquisitions (i.e., buying), management and conservation of such assets.

Core technical competencies include:

1 Object Identification;
2 Research Methodologies and Techniques;
3 Valuation.

Other areas of experience include technical competencies such as:

1 Auctioneering;
2 Capital Taxation;
3 Conservation and Restoration;
4 Insurance;
5 Purchase and Sale.

Planning and Development

Planning and Development Specialists will advise on the entire lifecycle of the built and natural environment, with a focus on planning legislation, development potential and sustainability.

Core technical competencies include.

1 Development Appraisals;
2 Planning and Development Management;
3 Spatial Planning Policy and Infrastructure;
4 Legal/Regulatory Compliance;
5 Valuation;
6 Measurement;
7 Surveying and Mapping.

Other areas of experience are wide ranging and include technical competencies such as:

1 Access and Rights Over Land;
2 Cadastre and Land Administration;
3 Compulsory Purchase and Compensation;
4 Contaminated Land;
5 Design and Specification;
6 Development/Project Briefs;
7 Economic Development;
8 Environmental Assessments;
9 Housing Strategy and Provision;
10 Master Planning and Urban Design;
11 Measurement;
12 Project Finance,
13 Purchase and Sale;
14 Risk Management.

Project Management

Project and Programme Managers are involved in construction projects from inception of the brief, to leading teams, procurement, tendering, advising on contracts and managing key stakeholders.

Core technical competencies include:

1 Contract Practice;
2 Development and Project Briefs;
3 Leading Projects, People and Teams;
4 Managing Projects;
5 Programming and Planning;
6 Construction Technology and Environmental Services;
7 Procurement and Tendering;
8 Project Finance.

Other areas of experience include technical competencies such as:

1 BIM Management;
2 Commercial Management;

3 Consultancy Services;
4 Contract Administration;
5 Development Appraisals;
6 Legal/Regulatory Compliance;
7 Maintenance Management;
8 Performance Management;
9 Purchase and Sale;
10 Stakeholder Management;
11 Supplier Management;
12 Works Progress and Quality Management.

Property Finance and Investment

Chartered Surveyors specialising in Property Finance and Investment may advise on both direct and indirect property investments. They are often employed in-house for banks or financial institutions or may work for fund or asset management firms.

Core technical competencies include:

1. Financial Modelling;
2. Inspection;
3. Investment Management (Including Fund and Portfolio Management);
4. Property Finance and Funding;
5. Valuation.

Other areas of experience include technical competencies such as:

1. Accounting Principles and Procedures;
2. Capital Taxation;
3. Corporate Finance;
4. Development Appraisals;
5. Indirect Investment Vehicles;
6. Landlord and Tenant;
7. Leasing/Letting;
8. Local Taxation/Assessment;
9. Property Management;
10. Purchase and Sale;
11. Research Methodologies and Techniques;

12 Strategic Real Estate Consultancy;
13 Valuation.

Quantity Surveying and Construction

Chartered Quantity Surveyors manage costs throughout the lifecycle of construction projects, including advice on feasibility, design and construction method. Quantity Surveyors can work on both the contractor and client-side in a variety of organisations.

Core technical competencies include:

1 Commercial Management (of Construction Works);
2 Design Economics and Cost Planning;
3 Construction Technology and Environmental Services;
4 Contract Practice;
5 Procurement and Tendering;
6 Project Finance (Control and Reporting);
7 Quantification and Costing (of Construction Works).

Other areas of experience include technical competencies such as:

1 Capital Allowances;
2 Contract Administration;
3 Corporate Recovery and Insolvency;
4 Due Diligence;
5 Insurance;
6 Programming and Planning;
7 Project Feasibility Analysis;
8 Risk management.

Research

Some Chartered Surveyors will be involved in Research relating to the delivery of solutions for projects. They are likely to specialise in one of the other pathways, with their research becoming the focus of the project concerned. This could include construction costs, financial modelling or investment analysis, for example.

Core technical competencies include:

1 Client Care;
2 Data Management;
3 Research Methodologies and Techniques.

Residential

Chartered Surveyors specialising in the Residential Sector, i.e., housing or domestic uses, may provide advice on a wide range of instructions, such as disposals, acquisitions, management, development, investment and valuation. Roles may be found across a wide range of organisations including in the private sector, block management, social housing, housing developers and local authorities.

There may be an element of cross over with the role of a Building Surveyor, particularly where a level 3 Building Survey is instructed as part of a residential sale or purchase instruction. Chartered Surveyors dealing with this type of work need a strong understanding of construction, building pathology and defects analysis.

For the AssocRICS assessment, there are three specific options within this pathway; Real Estate Agency, Residential Property Management and Residential Survey and Valuation.

Core technical competencies include:

1 Building Pathology;
2 Housing Maintenance, Repairs and Improvements;
3 Housing Management and Policy;
4 Housing Strategy and Provision;
5 Inspection;
6 Leasing and Letting;
7 Legal/Regulatory Compliance;
8 Market Appraisal;
9 Measurement;
10 Property Management;
11 Purchase and Sale;
12 Valuation.

Other areas of experience include technical competencies such as:

1 Auctioneering;
2 Capital Taxation;
3 Compulsory Purchase and Compensation;

4. Development Appraisals;
5. Landlord and Tenant;
6. Loan Security Valuation;
7. Local Taxation/Assessments;
8. Maintenance Management;
9. Property Management;
10. Purchase and Sale;
11. Strategic Real Estate Consultancy;
12. Supplier Management.

Rural

Chartered Surveyors who specialise in the Rural Sector will advise on a wide range of instructions, including disposals, acquisitions, management, valuation and planning. They may also have specialist knowledge of agriculture, auctioneering and rural support schemes.

Core technical competencies include:

1. Agriculture;
2. Management of the Natural Environment and Landscape;
3. Property Management;
4. Valuation.

Other areas of experience include technical competencies such as:

1. Access and Rights Over Land;
2. Auctioneering;
3. Compulsory Purchase and Compensation;
4. Forestry and Woodland Management;
5. Land Use and Diversification;
6. Landlord and Tenant;
7. Planning and Development Management;
8. Purchase and Sale.

Taxation Allowances

Chartered Surveyors may advise on Taxation relating to construction and existing land and property. This could include tax positions, tax allowances and tax optimisation.

Core technical competencies include:

1 Accounting Principles and Procedures;
2 Capital Allowances;
3 Construction Technology and Environmental Services;
4 Quantification and Costing of Construction;
5 Valuation;
6 Contract practice.

Other areas of experience include technical competencies such as:

1 Capital Taxation;
2 Contaminated Land;
3 Design Economics and Cost Planning;
4 Development Appraisals;
5 Due Diligence;
6 Insurance;
7 Property Finance and Funding;
8 Property Management;
9 Risk Management;
10 Sustainability.

Valuation

Valuers are involved with a wide variety of property assets and provide valuations for purposes such as secured lending, financial reporting, agency and internal decision making.

Core technical competencies involve other areas of experience taken from other pathways, such as Commercial Real Estate and Residential, but also include:

1 Inspection;
2 Valuation;
3 Measurement.

Valuation of Businesses and Intangible Assets

Business and Intangible Asset Valuers will provide advice on trading businesses, rather than solely property or land assets. This may involve transactional, taxational or financial advice, amongst many other potential reasons for valuations being provided.

Core technical competencies include:

1 Accounting Principles and Procedures;
2 Asset Identification and Assessment;
3 Valuation of Businesses and Intangible Assets;
4 Valuation Reporting and Research.

Other areas of experience include technical competencies such as:

1 Capital Allowances;
2 Compulsory Purchase and Compensation;
3 Corporate Finance;
4 Corporate Recovery and Insolvency;
5 Purchase and Sale;
6 Taxation.

Conclusion

In conclusion, Chartered Surveyors have a broad spectrum of roles available to them, in which they can gain experience. These are likely to change throughout a Chartered Surveyor's career and are not defined wholly by the AssocRICS or APC sector pathway pursued at the point of initial qualification.

Potential Chartered Surveyors should look for an area of professional practice that most interests them and offers attractive career prospects. That said, being open to opportunities and exploring new sectors or specialisms frequently offers exciting and rewarding new experiences.

In the next chapter, we will look at the routes to becoming a Chartered Surveyor. These provide appropriate options for a diverse range of candidates. We will also discuss the progression options available to experienced candidates, as well as new graduates, academics and those in senior professional positions.

3 Routes to Becoming a Chartered Surveyor

Introduction

We have looked at where you might find Chartered Surveyors working, including different sectors, industries and markets. In this chapter, we look at how you can become a Chartered Surveyor, i.e., what RICS route to qualification you should follow.

This doesn't just include the traditional graduate route for those who have completed an undergraduate degree course at university. There are also opportunities for senior professionals, academics, researchers, experienced surveyors, school leavers and candidates without any formal academic qualifications.

Becoming a Chartered Surveyor is entirely and practically possible for a diverse range of candidates, particularly those with good forward planning and commitment to achieving set goals. That said, there is often more than one possible route and deciding which to follow can be difficult.

This will become increasingly apparent as more school aged, school leavers and future professionals become more aware of the profession. T Levels and apprenticeships are becoming increasingly available as alternative qualifications. It is also relevant for mature, experienced candidates who want to become qualified and need to select a route which reflects their advanced role and responsibilities.

What Are the Routes to Becoming a Chartered Surveyor?

Simply put, there are many ways to become a Chartered Surveyor, which are represented in Figure 3.1 (RICS, 2019a).

DOI: 10.1201/9781003156673-3

Whilst some of the terminology used initially may seem alien, by the end of this chapter the terms should be a lot clearer. These concepts will be built on in the following chapters, which will look in much further detail at the AssocRICS and MRICS qualifications and assessment requirements.

First, we will look at T Levels and apprenticeships as routes to becoming both AssocRICS and MRICS qualified. There are many benefits to considering these routes to becoming a Chartered Surveyor, particularly as they allow candidates to gain practical work experience early on in their careers.

We will look at the routes to gaining the AssocRICS qualification, which can be used as a steppingstone to progress to MRICS at a future date. It is equally and highly valued in the industry as a standalone qualification and some AssocRICS surveyors may decide not to progress to MRICS. This is sometimes the case for residential surveyors who primarily undertake condition surveys and HomeBuyer Reports (with or without valuation).

We will look at degree level courses and the benefits of undertaking an RICS-accredited course. This may or may not include a placement year or year in industry. RICS-accredited courses typically lead to the Structured Training route, discussed later in this chapter. However, for experienced candidates undertaking an RICS-accredited degree course may lead straight to assessment without the need for any structured training. We will also briefly look at non-cognate qualifications, which provide access to the industry typically through the preliminary review process.

Below we will look at the specific routes providing access to the APC and MRICS qualification:

- APC – Structured Training, straight to assessment and preliminary review, depending on a candidate's experience and qualifications;
- Senior Professional Assessment;
- Specialist Assessment;
- Academic Assessment;
- Direct Entry;
- AssocRICS Progression.

Importantly, candidates need to be aware that the RICS operate a 'six-year rule'. This means that from the date of enrolment, all

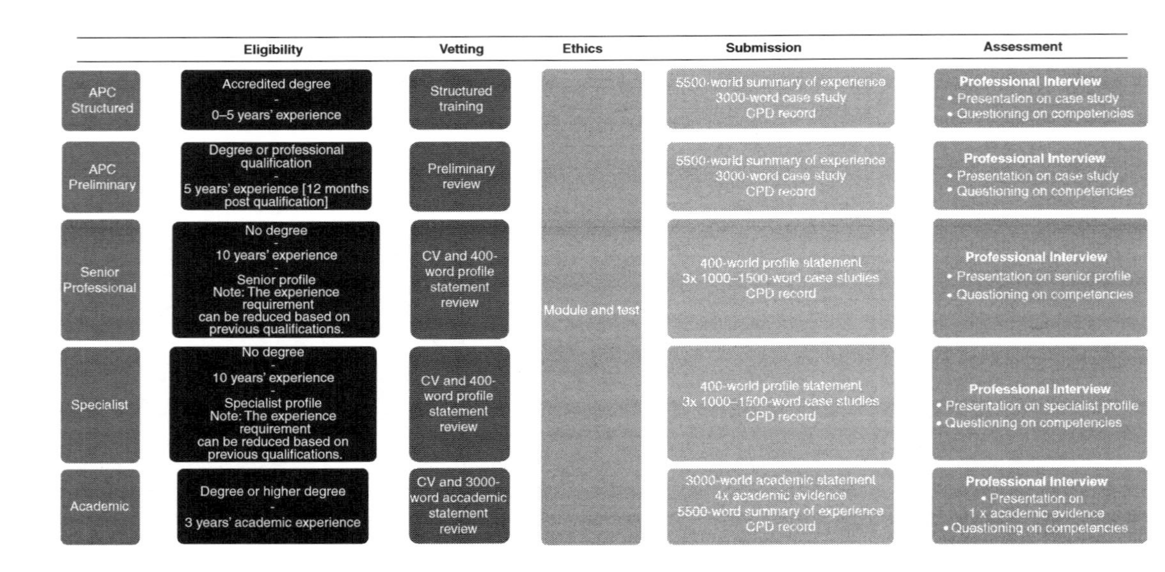

Figure 3.1 Routes to Becoming a Chartered Surveyor

candidates have six years to achieve either the AssocRICS or MRICS qualifications. If a candidate achieves AssocRICS, then the 'six-year rule' requirement is mitigated in relation to attaining MRICS status.

T Levels

T levels were launched in September 2020, providing a viable, vocational alternative to A levels. They are aligned in accordance with the needs of specific industries and sectors. Their structure includes a significant element of work-based experience.

T levels sit between two other common post-GCSE options; A levels, which provide academic education, and apprenticeships providing even more industry work experience, i.e., 80% 'on the job training', supplemented by 20% classroom-based learning.

A T level takes two years to complete and equates to three A levels. It blends traditional classroom learning with industry experience, equating to at least 315 hours, or 45 days. This industry experience can be gained through a variety of modes, including day-release, blocks, or a mix of the two.

By 2023, T levels will be provided in twenty-four subjects and eleven industry areas, including accounting, catering, digital business services, finance, health and legal.

For candidates wishing to become a Chartered Surveyor, the following T levels will be of relevance:

- Design, Surveying and Planning for Construction;
- Onsite Construction (launched 2021).

The T level in Design, Surveying and Planning for Construction includes modules on:

- Health and Safety;
- Science;
- Design;
- Construction and the Built Environment Industry;
- Sustainability;
- Measurement;
- Building Technology;

- Information and Data;
- Relationship Management;
- Digital Technology;
- Commercial and Business.

There are four further occupational specialisms within this T level:

- Surveying and Design for Construction and the Built Environment;
- Civil Engineering;
- Building Services Design;
- Hazardous Materials Analysis and Surveying.

Successful T level candidates will be awarded a grade of pass, merit, distinction or distinction*, together with separate grades for the core component (E to A*) and occupational specialism (pass, merit or distinction).

If a T level candidate subsequently wishes to apply for an undergraduate university course, then they will be able to use UCAS tariff points allocated to their overall T level grade (Gov.uk, 2020):

Table 3.1 UCAS Tariff Points for Overall T Level Grades

UCAS tariff points	*T Level overall grade*	*A level*
168	Distinction* (A* on the core and distinction in the occupational specialism)	AAA*
144	Distinction	AAA
120	Merit	BBB
96	Pass (C or above on the core)	CCC
72	Pass (D or E on the core)	DDD

In summary, T levels provide an excellent way to start a career in surveying. They provide a holistic overview of the industry with an element of work experience. This can then be consolidated through further education, qualifications or experience to become AssocRICS or MRICS qualified in the future.

Apprenticeships

Apprenticeships provide 'on the job' training and learning how to become a Chartered Surveyor. This includes the delivery of an academic qualification by a university, for example, and paid employment within the industry by an appropriate firm or organisation. Apprenticeships are available to candidates over 16 years of age, although there is no requirement to be a school leaver and there is no upper age limit.

The way that apprenticeships are funded in England means that apprentices do not need a student loan to pay for their university course. There are different processes and funding in place in Scotland, Wales and Northern Ireland.

There are different types and levels of surveying apprenticeship currently available.

Level 3 (A level equivalent – two years) Surveying Technician, where the end point assessment is to become AssocRICS qualified.

The following pathways are available:

- Building Surveying;
- Commercial Property;
- Land;
- Planning and Development;
- Project Management, Residential Property;
- Valuation;
- Quantity Surveying.

Level 6 (degree equivalent) Chartered Surveyor, where apprentices complete an RICS-accredited undergraduate degree following sixty months' study, a PGDip or a master's degree following thirty months' study (if any undergraduate degree is already held). The end point assessment is the APC to become MRICS qualified.

The following pathways are available:

- Building Surveying;
- Quantity Surveying and Project Management;
- Property, which aligns to several APC pathways including Corporate Real Estate, Commercial Real Estate, Land and Resources, Planning and Development, Residential, Rural and Valuation.

As the Level 6 Chartered Surveyor apprenticeship includes RICS-accredited postgraduate courses, not only undergraduate courses, but non-cognate degree holders can also pursue an apprenticeship to become a Chartered Surveyor. This has widened entry into the industry for an even more diverse range of candidates, particularly those who are re-training or changing careers later in their working lives.

Level 3 (A level equivalent – two years) Geospatial Survey Technician. After completion, candidates can apply via AssocRICS direct entry to become AssocRICS.

Level 6 (degree equivalent – five years) Geospatial Mapping and Sciences. After completion, candidates can apply via APC direct entry to become MRICS.

As a side note, T levels and apprenticeships may sound similar on paper. However, T level subjects are much broader and relate to wider industries or sectors. They also provide more overall pastoral and educational support given their primary, classroom-based mode of learning. In contrast, apprenticeships tend to provide a lower level of pastoral support as the apprentice is primarily based in the workplace and supervised by their employer, rather than the education provider.

Apprenticeships, therefore, provide an ideal option for focussed candidates who are able and committed to self-manage their time and learning. This may particularly be attractive to more mature candidates who want to learn whilst being 'on the job'.

AssocRICS

The AssocRICS qualification is based on a written submission only.

It is open to candidates who have the following:

- One year of relevant experience and a relevant undergraduate degree;
- Two years of relevant experience and a relevant higher, advanced or foundation level qualification;
- Four years of relevant experience and no academic qualifications.

AssocRICS status demonstrates to clients, peers and the public that a surveyor is competent to undertake their role. It does not provide Chartered status but can provide a platform to achieve this if a candidate wishes to do so.

University Courses

Access to becoming a Chartered Surveyor is often facilitated by undertaking an RICS-accredited degree. This will be discussed further in the next section explaining the various APC routes to becoming MRICS qualified. Essentially, pursuing an RICS-accredited degree means that candidates do not have to undergo preliminary review, i.e., an additional assessed written assessment. This can reduce the minimum amount of experience that is required to be eligible to apply for the APC final assessment interview.

Generally, RICS-accredited courses require approximately 120 UCAS tariff points, depending on the institution and course. Other requirements include a GCSE Grade C/4 in English Literature, or Language, and in Mathematics. A wide range of A level subjects are accepted, although subjects such as Business, Economics, Geography, Law and Mathematics may be particularly useful to gain baseline knowledge relevant to surveying courses.

The RICS provide a dedicated website (RICS, 2020g) listing over five hundred accredited courses internationally. Approximately three hundred of these are provided by UK institutions. Examples of typical courses include Building Surveying, Construction Project Management, Property Development and Planning, Quantity Surveying and Commercial Management and Real Estate.

Many courses will offer a year in industry or a placement, or sandwich year. These are an excellent way for students to gain practical work experience, attaining the first 12 months or 200 days of their APC structured training. Candidates will then need to undertake a further 12 months, or 200 days of structured training after graduating to satisfy the minimum RICS requirements. Of course, these are only the minimum requirements and many candidates will decide to accrue more experience prior to submitting their final assessment documents.

Other advantages to consider are the many different modes of course attendance, including full-time, part-time, day-release and distance learning. This means that many candidates can work alongside undertaking a degree course. This decision should not be taken lightly, and candidates should be aware of the significant time requirements of any degree course. This is similar to an apprenticeship, but without the same government funding benefits, or the level of support from their employer and academic course provider.

APC and MRICS

The APC is the assessment process that candidates must undergo to become MRICS qualified, i.e., to become a Chartered Surveyor. The assessment process includes a written submission and a face-to-face, online interview.

There are various options available to undertake the APC, including structured training, preliminary review in the form of an additional written assessment, or straight to assessment. These are generally appropriate for candidates working in the surveying industry with technical day-to-day responsibilities.

The APC also has three separate and distinct assessment routes for candidates that are in senior professional, specialist or academic positions. These assessments are very different to the 'traditional' APC routes outlined above.

Looking at the 'traditional' assessment routes first, the following options are available depending on a candidate's experience and academic qualifications:

- Structured Training (24 months/400 days) – RICS-accredited degree and no prior experience, usually pursued by recent graduates;
- Structured Training (12 months/200 days) – RICS-accredited degree and at least five years of relevant experience, this route is typically pursued by more experienced candidates or those who have undertaken a longer length part-time, day-release or distance-learning degree course;
- Straight to Assessment – RICS-accredited degree and at least ten years of relevant experience, generally pursued by experienced candidates who are not in senior professional, specialist or academic positions;
- Preliminary Review – Non-cognate degree, either undergraduate or postgraduate, or membership of an RICS approved professional body (see Direct Entry section below for further explanation) and at least five years of relevant experience, of which twelve months must have been post-graduation.

This chapter will now look in turn at each of the 'traditional' routes to APC assessment, before turning to the senior professional, specialist and academic assessments.

Structured Training

Structured Training is appropriate for candidates with an RICS-accredited degree and up to ten years of experience, as outlined above. For candidates with five to ten years of relevant experience, pre-qualification experience can count towards the minimum RICS requirements.

Candidates will need to work with their employer to ensure that their training is structured to meet the requirements of their chosen sector pathway, e.g., Building Surveying or Project Management and relevant competencies. The minimum period of structured training will either be 24 months, encompassing 400 days, or 12 months of 200 days depending on the candidate's prior experience, as detailed above. These are the minimum requirements, however, and many candidates will continue accruing experience until they feel ready and confident to submit for final assessment.

The earliest candidates undertaking structured training can apply for final assessment is at the end of month 11 or 23 and then sit their final assessment after month 12 or 24. This is dependent on whether their structured training period is for 12 or 24 months.

A candidate's structured training will be logged via their diary and accessed via the online RICS Assessment Resource Centre (ARC). Candidates will also need to meet at least every three months with their Counsellor, and Supervisor, if appointed, to keep track of their progress and identify any gaps in knowledge or experience.

The RICS provide a downloadable Candidate Training Plan (RICS, 2020h), which can be used to plan a candidate's period of structured training. This creates a clear commitment and agrees expectations between the employer and the candidate, in addition to establishing clear goals for the candidate to achieve.

Straight to Assessment

Some candidates will be able to proceed straight to assessment without the need for any structured training or preliminary review submission. This is generally for candidates with an RICS-accredited degree and at least ten years of relevant experience.

The RICS can provide candidates with a self-assessment form to identify any gaps in their experience or competencies. Ideally, three monthly meetings will be held with their counsellor, and supervisor if appointed. The intention of these meetings is to keep progress on track.

Preliminary Review

Candidates who do not have an RICS-accredited degree, i.e., a non-cognate degree, will need to pursue the preliminary review route.

This essentially means that a candidate prepares their submission and submits this for a written review by the RICS. This is submitted by the candidate before their final assessment, thus allowing the RICS to check that a candidate has attained the appropriate level of professionalism, knowledge and experience. This mitigates the requirement for the candidate to hold an RICS-accredited degree, where the RICS will already have 'pre-approved' the course structure and content.

If a candidate's preliminary review submission is acceptable, then they may submit their written documents in a future final assessment submission window. This is generally four to six months later, but a candidate can wait until a subsequent submission assessment window if they wish, e.g., if they are unusually busy at work or have family commitments which need to come first.

If a candidate's preliminary review submission is not acceptable, then they will receive a feedback report from the RICS and will need to re-submit for preliminary review, before they can proceed to submit for final assessment. Only when the preliminary review stage has successfully been passed can a candidate proceed to present their final written assessment and proceed to the interview stage.

Senior Professional

The Senior Professional Assessment is one of three separate APC assessment routes. This requires candidates to have at least ten years of relevant experience, which is reduced to five years if a postgraduate degree is held. Candidates also need to be able to demonstrate senior professional responsibilities in their role, including leadership, management of people and management of resources.

The Senior Professional Assessment is often more appropriate for candidates in senior management positions or who are responsible for managing or leading teams, rather than being responsible for day-to-day technical surveying work. Indicators of this assessment route being appropriate are a candidate's senior position in their organisation's structure, advance decision-making responsibilities, an international dimension to a candidate's role, high profile client base and recognition from the wider industry, peers and media. A candidate does not need to satisfy all of these indicators, but a number should be present, e.g., they may have a senior role and a high-profile client base, but no international dimension.

The structure of the senior professional written final assessment and interview are different to the 'traditional route'. The emphasis is on the three senior professional competencies of leadership, management of people and management of resources.

Specialist

As with the Senior Professional Assessment, the Specialist Assessment requires candidates to have at least ten years of relevant experience, reduced to five years if a postgraduate degree is held. Candidates also need to have advanced responsibilities for a specialist or niche area of work.

Specialists may have a high-level decision-making position, a track record of providing specialist consultancy work, having spoken at conferences, written in the trade press, been appointed by a governance or judicial body, been recognised by the wider industry, peers and media. They may have lectured, provided formal training, be qualified above master's level, or involved in dispute resolution for a specialist technical area. Again, specialists do not need to demonstrate all of these indicators, but several should be demonstrated.

The structure of the specialist written assessment and interview are different to the 'traditional' route, with an emphasis on the candidate's specialist area of expertise.

Academic

The Academic Assessment is appropriate for academic professionals, e.g., lecturers or researchers, who have at least three years of academic experience and a surveying-related degree. The academic experience does not have to be continuous and can have been undertaken at different points in time. Experience can be drawn from teaching, research, scholarship, external engagement and academic activities.

The structure of the academic written assessment and interview are different to the 'traditional' route, with strong evidence of the candidate's academic experience required.

Direct Entry

Direct Entry provides access to the AssocRICS and MRICS qualifications for candidates who already hold RICS approved qualifications.

The Direct Entry Route requires candidates to submit proof of their approved qualification or professional body membership, a letter of good standing from the professional body, their recent Continued Professional Development (CPD) record. They also require support from an existing MRICS or FRICS Chartered Surveyor for APC Direct Entry. For AssocRICS Direct Entry, an existing MRICS or FRICS Chartered Surveyor or an AssocRICS member with at least four years of post-membership experience can support a candidate's application.

At AssocRICS, examples include being a Member (MCIAT) of the Chartered Institute of Architectural Technologists (CIAT), Member (CIHCM) of the Chartered Institution of Housing (CIH) or a Member (MIRPM) of the Institute of Residential Property Management (IRPM).

For candidates following the Residential Survey and Valuation pathway, a common direct entry option is via the Surveyors and Valuers Association (SAVA) Diploma in Residential Surveying and Valuation, plus accruing at least two years of experience. There are also a variety of global qualifications and memberships, e.g., being a Full Member of the Hong Kong Institute of Clerks of Works (HKICW).

At MRICS, there are two levels of Direct Entry; no requirement to undertake the APC (i.e., MRICS without the need for any further assessment) and APC assessment via preliminary review, Senior Professional Assessment or Specialist Assessment. All direct entry candidates need to complete the RICS online ethics module and test.

In the UK, qualifications which allow candidates to become MRICS without any further assessment, i.e., there is no requirement to complete the APC assessment, include being a Professional Member (MICFor) of the Institute of Chartered Foresters (plus holding a relevant undergraduate degree and ten years of post-membership experience) or a Member (MCInstCES) of the Chartered Institution of Civil Engineering

Surveyors (ICES) (plus holding a relevant undergraduate degree and five years of post-membership experience, or ten years of post-membership experience and no degree).

Globally, other qualifications include being a Licensed Land Surveyor in New Zealand with at least five years of experience, or a Professional Quantity Surveyor (PrQS) within the South African Council for the Quantity Surveying Profession (SACQSP).

Other qualifications allow Direct Entry candidates to qualify as MRICS via the APC preliminary review process, Senior Professional Assessment or Specialist Assessment. These include being a Member (MACostE) of the Association of Cost Engineers, a Member (C. Build E MCABE) of the Chartered Association of Building Engineers, a Member (MCIOB) of the Chartered Institute of Building (CIOB), Chartered Member of the Institution of Structural Engineers (ISE) or a Chartered Town Planner, i.e., a Member (MRTPI) of the Royal Town Planning Institute (RTPI). Again, there are various global qualifications such as being a Full Member of the Shanghai Construction Cost Association (SCCA) in China, or Full Member of the Society of Industrial and Office Realtors (SIOR) in the USA.

AssocRICS Progression

AssocRICS surveyors can pursue MRICS qualification via several different APC routes. This depends on the candidate's existing experience, academic qualifications and seniority.

AssocRICS candidates who already have a degree (RICS-accredited or not) or approved professional body membership can pursue their APC via the Structured Training or Preliminary Review Routes. AssocRICS qualified candidates can also pursue the Senior Professional, Specialist and Academic Routes if these routes are appropriate for their role and responsibilities.

There will also be AssocRICS candidates who have no degree or approved professional body membership. When four years of post-AssocRICS experience has been accrued, candidates can proceed via the AssocRICS progression route. The last year of this should follow a Structured Training approach, although if this is not possible then candidates can instead proceed via the Preliminary Review Route.

The AssocRICS Progression Route requires 900 study hours to be undertaken from the final level, i.e., the final year of an RICS-accredited undergraduate, or postgraduate degree course. The chosen modules should be relevant to the candidate's chosen pathway and competencies and will typically be undertaken in 150, 200 or 300 hour blocks. Candidates should contact higher education institution providers to identify suitable RICS-accredited course and module options.

Out of the 900 study hours, 300 may be satisfied via work-based learning, a minimum of two years for repeated tasks, structured in-house training or private study (not necessarily from an RICS-accredited qualification). This must be approved by the RICS and be accumulated in blocks of 150 hours plus.

Candidates can then proceed to submit their final written APC submission and sit their online assessment interview.

There may be more than one available route for AssocRICS surveyors to consider when proceeding to become MRICS, so careful consideration should be given to which meets the candidate's individual requirements. This could be based on timing requirements, knowledge gaps or employer support, for example.

Conclusion

In conclusion, there are various routes to becoming a Chartered Surveyor catering for candidates from a wide range of backgrounds, experience and qualifications. This includes new entrants to the industry at school leaving age, all the way through to non-cognate, mature and highly experienced professionals.

In the next chapter, the APC assessment process will be discussed in detail, including the nature and key requirements of the written assessment and online assessment interview.

4 APC

Introduction

In the last chapter, we looked at the various APC routes to becoming a MRICS Chartered Surveyor. In this chapter, we look specifically at the APC process and structure of both the written submission and online interview.

This will explain the preliminary review submission, final assessment submission and final assessment interview. It will then outline the specific requirements of the senior professional, specialist and academic assessments. All APC assessments include two elements: a written submission and an online final assessment interview.

Enrolment

All candidates need to enrol on the APC as a starting point via the RICS website. This will require the candidate to have chosen the most appropriate APC route for their individual requirements, as outlined in Chapter 3. Successful enrolment will allow the candidate access to the online RICS Assessment Resource Centre (ARC), which will be discussed later in this chapter.

After enrolment, candidates will need to set up their profile on ARC, which will be discussed later in this chapter. This requires the candidate to confirm their pathway and competency choices. These form the basis of each individual candidate's written submission and should be selected before a candidate starts drafting any written work.

All candidates will need an allocated Counsellor, a role which will be discussed in a later chapter. Essentially, however, this is someone

DOI: 10.1201/9781003156673-4

who knows the candidate well who will support them through the APC process and sign off their submission before they can proceed to the final online interview stage.

Candidates should also engage their employer in their APC process and may identify a Supervisor to provide additional support, alongside their Counsellor. It is only a mandatory requirement to have an allocated Counsellor, however. Importantly, candidates should feel supported throughout their APC journey, but the onus is on the candidate's motivation to drive forward and be committed to their own APC progress and achievement.

Candidates seeking to apply for the senior professional, specialist or academic assessments will need to first undergo a vetting procedure to ensure that the route is right for their knowledge, experience and role. This will also be discussed in the relevant sections later in this chapter.

Candidates enrolling on the APC structured training route can backdate the start of their structured training period by up to one month. This requires the candidate to notify the RICS by email, together with a supporting letter from the candidate's Counsellor or employer.

Candidates undertaking work placements, placement years or sandwich years can enrol on the structured training route during this period. This requires the candidate to register as an RICS Student member and to notify the RICS by email that they are beginning to record experience towards their minimum structured training requirement.

At the end of a candidate's placement or work experience, the candidate will need to obtain written confirmation from their Counsellor, an MRICS or FRICS Chartered Surveyor, of the competencies and number of days of experience that have been achieved. The candidate will then be able to formally enrol on the APC structured training route upon starting employment post-graduation, needing to complete only 12 months of further structured training to meet the minimum RICS requirements. It is often advisable for candidates to undertake more than just the minimum requirements to gain the appropriate level of knowledge, experience and confidence before proceeding to final assessment.

Candidates will also have to pay an enrolment fee to the RICS, which covers the enrolment process, ethics module and test and final assessment interview. The enrolment fee is slightly increased for preliminary review candidates. All candidates must pay a yearly subscription fee covering access to ARC and all supporting materials provided by the RICS.

From the point of enrolment, all candidates commit to qualifying within a six year period, known as the six year rule. As previously discussed, this may be mitigated by qualifying as AssocRICS as a stepping stone to becoming MRICS via the APC at a later date.

Overview of the APC Assessment

All APC candidates, irrespective of their chosen route, must undergo assessment via a written submission and an online interview. However, the requirements of both elements differ significantly for the senior professional, specialist and academic assessments, which will be considered separately in this chapter.

The written submission element is also slightly different for preliminary review candidates. This is because preliminary review introduces a two-stage written submission process, which will be discussed separately to the final assessment submission. The requirement to undergo preliminary review is set out clearly in the previous chapter.

The online interview is based on the candidate's written submission and lasts for one hour. The final decision as to whether a candidate becomes MRICS is solely based on performance in this interview. There is no separate assessment of the written submission, apart from the requirement for preliminary review candidates to pass this element of their written submission process and for final assessment candidates to meet the minimum RICS requirements, e.g., word count and case study date validity.

What Are the APC Pathways?

There are twenty-two sector pathways which APC candidates can choose from, which have already been discussed at length. All candidates, including those undertaking the senior professional, specialist and academic assessments, must choose a sector pathway which most closely reflects their experience, role and knowledge:

1. Building Control;

2. Building Surveying;
3. Commercial Real Estate;
4. Corporate Real Estate;
5. Environmental Surveying;
6. Facilities Management;
7. Geomatics;
8. Infrastructure;
9. Land and Resources;
10. Management Consultancy;
11. Minerals and Waste Management;
12. Personal Property/Arts and Antiques;
13. Planning and Development;
14. Project Management;
15. Property Finance and Investment;
16. Quantity Surveying and Construction;
17. Research;
18. Residential;
19. Rural;
20. Taxation Allowances;
21. Valuation;
22. Valuation of Businesses and Intangible Assets.

What Are the APC Competencies?

Competencies are the individual mandatory ('soft') and technical ('hard') areas where candidates need to demonstrate their skills, knowledge and experience. The competency requirements for each separate pathway are set out in the relevant pathway guides, for example, Corporate Real Estate, which can be downloaded from the RICS website.

The requirements of each specific competency are divided into three levels. The level of competence (either levels 1, 2 or 3) required by a candidate will depend on their chosen pathway and competencies. For example, all candidates must satisfy the mandatory competency, Ethics, Rules of Conduct and Professionalism to level 3. The pathway guides set out the individual requirements of each level and each competency, e.g. required knowledge for level 1 and required activities or advice at levels 2 and 3 respectively.

Candidates should read the competency descriptors in each relevant pathway guide very carefully to ensure that they choose competencies which reflect their knowledge and experience. Candidates should continually refer to the competency descriptors to ensure that what they write in their submission is accurate, relevant and sufficiently detailed. This is because what is written may lead to potential areas of questioning in a candidate's final assessment interview.

The three levels are defined as follows:

- Level 1 – Knowledge and Understanding – This requires candidates to explain their knowledge and learning relevant to the description included in the competency guide. This could be through academic learning, e.g., degree level course, CPD or on-the-job training. However, candidates should avoid too much repetition as their CPD record will be set out separately. Generally, level 1 will be met by including a brief list of knowledge or relevant topics, given the limitations of the overall word count, which will be discussed later in this chapter.
- Level 2 – Application of Knowledge and Understanding – Demonstrating level 2 relates to a candidate applying their level 1 knowledge and understanding in practice, i.e., through practical work experience. This relates to 'doing', rather than just knowing the theory and fundamentals, or underpinning theory with practice. Candidates should refer to specific projects, instructions or examples to clearly demonstrate their practical activities and experience, including their role and relevance to the competency description.
- Level 3 – Reasoned Advice and Depth of Knowledge – Expanding on level 2 work experience or tasks to provide reasoned advice to clients. This involves considering the options available and offering recommendations and advice on solutions. Candidates should ensure that they explain their role in the first person, using I, me and my. Candidates should avoid referring to collective involvement such as we, our or us.

Every pathway has the same mandatory competencies, as well as technical competencies which are split into two categories; core (primary skills) and optional (additional skills).

What Are the Mandatory Competencies?

The eleven mandatory competencies for all candidates are 'soft skills' or business skills, common to all pathways. This means that they need to be demonstrated by all APC candidates, irrespective of role, sector or position. They aim to show that a candidate is able to work with others, can manage their own workload and can act ethically and professionally.

The mandatory competencies include:

- Ethics, Rules of Conduct and Professionalism (level 3);
- Client Care (level 2);
- Communication and Negotiation (level 2);
- Health and Safety (level 2);
- Account Principles and Procedures (level 1);
- Business Planning (level 1);
- Conflict Avoidance, Management and Dispute Resolution Procedures (level 1);
- Data Management (level 1);
- Diversity, Inclusion and Teamworking (level 1);
- Inclusive Environments (level 1);
- Sustainability (level 1).

Candidates are also able to pursue some of the mandatory competencies to higher levels as part of their core, or optional competency choices on certain pathways.

What Are the Technical Competencies?

The technical competencies are the 'hard' skills required of candidates by the RICS. They are split into core and optional competencies, depending on the specific requirements of each pathway. Candidates should pay particular attention to the requirements of the core competencies as these must be demonstrated to the required level. This may require candidates to seek additional work experience or secondment to fill gaps in experience, e.g., Planning and Development pathway candidates may need to identify additional Valuation experience to fulfil this as a core technical competency to the required level.

Final Assessment Submission (excluding Senior Professional, Specialist and Academic Assessments)

All APC candidates, whether undergoing preliminary review or not, will need to submit a written final assessment submission. This applies to candidates on the structured training and straight to assessment routes, as well as candidates who have successfully passed preliminary review. The final assessment submission is, therefore, exactly the same for all of these routes. The final assessment submission is very different in structure and content for the senior professional, specialist or academic assessments, however.

The written submission will be the product of a substantial period of preparation and not something that can be drafted in a matter of weeks. Rushing the written submission is likely to lead to a poor-quality document, which will not support the candidate through their online interview. It is vitally important for candidates to ensure that they present their submission to the highest written standards, including spelling, grammar and proofreading. Therefore, the final submitted documents need to be 'client ready' and written in formal, professional language.

The APC final assessment submission includes the following elements:

- Diary (structured training candidates only);
- Summary of Experience;
- Case Study;
- CPD Record;
- Ethics Module and Test.

Throughout their written work, candidates should be acutely aware of plagiarism. Under no circumstances should a candidate seek to include work or text written by another candidate or copied from a published source, unless it is correctly and clearly referenced. Any red flags of plagiarism will be identified by the RICS using the Turnitin plagiarism detection system, leading to further investigation and potential disciplinary action, including removal from the APC assessment process.

What Does the Structured Training Diary Include and Who Needs to Complete it?

The diary is only required for candidates on the structured training route. Candidates undergoing preliminary review or proceeding straight to assessment do not need to keep a diary.

Structured training candidates should start to record their diary via ARC as soon as they enrol on their APC journey. This is because the number of diary days logged counts towards satisfying the minimum twelve or twenty-four month structured training requirements. Candidates will also use the diary when compiling their summary of experience, choosing their case study topic and for discussion of their progress during Counsellor meetings.

Candidates are only required to log time spent on their technical competencies in their diary. This means that they should not record any activities or time spent on their mandatory competencies, unless these are also selected as technical competencies Mandatory competencies will generally be included within a candidate's daily work rather than being their specific focus at any one time. Examples include communicating during the course of an instruction or working within a project team.

Time spent on the mandatory competencies, therefore, will not contribute towards a candidate's minimum structured training requirements. However, they must still be recorded in the summary of experience, which is discussed later in this chapter.

Sufficient detail should be included in a candidate's diary, allowing appropriate examples to be identified to meet the requirements of the chosen competencies at levels 2 and 3. The diary does not form part of the candidate's final written submission, although the RICS may request it to be submitted separately prior to a candidate's online interview. Reasons for this include where the RICS identify anything concerning or contentious within a candidate's submission, e.g., the case study topic does not reflect the candidate's declared competencies or the standard of work in their summary of experience.

Therefore, candidates should ensure that the detail in their diary is sufficiently relevant and concise, rather than being extremely detailed and verbose. It should be written to a high standard of English and in a professional style.

The following details need to be recorded in ARC for each separate diary entry:

- Competency and level relevant to the diary entry;
- Days and start date, logged in a minimum of half day blocks;
- Title of the activity;
- Diary entry, including a brief description of the activity and relevant competencies. This should be as specific as possible to

aid the candidate's memory when they come to draft their written submission or for discussion with their Counsellor.

Candidates can combine their activities or experience within their diary entries into more substantial blocks of time. This may be sensible given the wide range of activities that a candidate is likely to undertake in the space of a day or week, for example. This might mean that a candidate decides to block together measurement or inspection activities from a variety of different projects or instructions into one diary entry. A candidate may also decide to block together time spent on one larger project into one entry to avoid repetition. How a candidate decides to split their entries is likely to be directed by their competency choices, as some are much wider topic areas than others. Valuation, for example is typically much wider than Measurement.

ARC provides a useful filter function by month, year, competency and level. The diary entries will also be shown on the individual competency pages on ARC, so it is important to ensure that the detail is accurate and relevant. This will facilitate Counsellor meetings and provide a very useful aide memoire when a candidate begins to draft their written submission.

What Is the Summary of Experience?

The Summary of Experience focusses on the candidate's competencies. This includes the mandatory competencies relevant to all candidates, as well as the technical competencies specific to a candidate's chosen pathway.

The technical competencies are split into core technical competencies, i.e., a requirement for the specific pathway, and optional technical competencies, i.e., where the candidate can select competencies which best suit their role, knowledge and experience. The chosen competencies will either have a stated achievement level, i.e. Ethics, Rules of Conduct and Professionalism to level 3, or candidates may have a choice of the level attained, e.g. optional technical competencies.

The word counts for the summary of experience are absolute; 1,500 for the mandatory competencies and 4,000 for the technical (core and optional) competencies. There is no leeway to exceed these and candidates can be referred for exceeding them before they reach the interview stage.

The word count for each section (i.e. mandatory and technical competencies) can be allocated as the candidate wishes between each separate competency. Generally, it is advisable to allocate a higher word count to the level 3 statements to ensure that sufficient experience and detail can be included.

The summary of experience requires candidates to write a summary of their knowledge at level 1 and experience at levels 2 and 3 for each level of each chosen competency. Unlike AssocRICS, this includes Ethics, Rules of Conduct and Professionalism. Candidates should ensure that they refer directly to the pathway guide and competency descriptors when preparing their summary of experience, as this is what they will be assessed against by their final interview panel.

If a competency is selected to level 3, then candidates must write a statement for each of levels 1, 2 and 3. For level 2, a statement must be provided for levels 1 and 2. At level 1, only a level 1 statement is required. At levels 2 and 3, it is advisable to use sub-headings or other appropriate formatting to highlight examples, aiding the candidate and assessment panel in the final assessment interview by sign posting specific examples.

At level 1, candidates must explain what they know or have learnt about a competency. This may link to CPD activities or academic learning, e.g., university modules. Given the limited word count, a bullet point list of knowledge could be used to reflect the competency descriptor, although a carbon copy of this is not allowable and would constitute plagiarism.

Candidates should ensure that the knowledge they declare is relevant to their experience, as anything included in the written submission can be questioned by the assessment panel. Candidates should ensure in level 1 that they demonstrate current knowledge of legislation and RICS guidance, together with any relevant hot topics.

At level 2, candidates should include two to three practical examples of their work-based experience. This will discuss how they acted or carried out relevant tasks. The range of relevant activities or tasks for a specific competency should be related to the pathway guide and competency descriptor, although these are not exhaustive lists.

Candidates should seek to be as specific and refined as possible in their examples. This could be by stating a specific instruction, property or client, to help focus their final interview questioning. Candidates should avoid using vague or broad examples as this can make it harder to demonstrate the required level in the final assessment interview. If a project or instruction is very broad, then candidates could focus on a specific aspect of their involvement or advice which clearly meets the requirements of the relevant competency; for example, measurement of a property as part of a valuation instruction, or inspection in the wider context of giving planning and development advice.

At level 3, candidates should again include two to three practical examples of their work-based experience. These examples should relate to where the candidate has given reasoned advice to a client. Again, candidates should use very specific examples, which clearly demonstrate their integral role in a project or instruction.

Candidates should ensure that they write any examples using the first person, i.e. using I, me and my. This ensures that the assessment panel know that the work is the candidate's rather than that of a colleague or superior. This also provides the assessment panel with the perception that the candidate has given professional advice and handled instructions competently from start to finish, with minimal supervision.

What Is the Case Study?

The case study provides an in-depth discussion of a specific project or projects that a candidate has been involved in and provided reasoned advice in relation to. Candidates should ideally focus only on one project, which helps to keep the detail refined and the project easy to understand by the assessment panel. The choice of project or instruction may include work outside of the candidate's geographical assessment region. However, in the final assessment interview, candidates should be aware of the legislation and guidance relating to both the geographical regions of the case study project and assessment location.

The choice of project should reflect the candidate's chosen pathway and competencies. However, the case study does not need to include aspects of every single competency, which is unlikely to be realistically achievable. The competencies demonstrated should include level 3 aspects, clearly demonstrating giving reasoned advice to a client.

The candidate must have been involved in the case study project or instruction within the last two years, from the date of submission. If a project is out of date, then this is a point for referral and a candidate will not be permitted to proceed to the online interview stage. Some projects, e.g., a construction project or a site disposal, may extend over a period of more than two years. However, the candidate's main involvement must have taken place during the two year validity period. It is helpful to include a clear timeline to demonstrate the candidate's involvement in their case study project. This can also help to demonstrate a candidate's visual communication skills.

The candidate should have played a key role in the case study project. For some instructions, this will include running, managing and leading the project from start to finish. For larger, more complex projects, a candidate may not have been running the project independently but will have played a key role in giving reasoned advice to the client.

Some candidates may have been involved in only one aspect of a project, been involved after the commencement of a project or have finished their involvement before the project completed. This is acceptable, as long as the candidate is able to demonstrate their reasoned advice in the stated level 3 competencies. If a project has not completed at the time a candidate writes their case study, then they may wish to discuss the prognosis within the case study and provide an update in their final interview presentation.

The case study choice does not need to be the most complex, high value or large project or instruction that the candidate has been involved with. Given the relatively limited word count, it can be challenging and counterproductive to use a very complex case study which the assessment panel do not understand clearly or fully. This can lead to misunderstandings and confused questioning as the candidate's role and advice has not been made clear in the context of the limited word count and written case study structure. The best case studies are often simple but succeed in setting out a clear and logical story of the candidate's involvement and reasoned advice.

Within their project or instruction choice, candidates should identify and discuss two or three key issues, involving challenges they encountered and needed to overcome during the course of the instruction. This will demonstrate the candidate's problem solving and analytical skills, leading to the provision of reasoned advice and recommendations to the client. If a candidate is unable to identify at least two key issues, then the project or instruction is unlikely to be a suitable case study topic.

Example case study key issues for a secured lending valuation instruction could include:

- Key issue 1 – Assessment of Market Rent;
- Key issue 2 – Assessment of Market Value, including choice of yield;
- Key issue 3 – Advice on secured lending to the client.

Example case study key issues for a level 3 Building Survey could include:

- Key issue 1 – Defect 1 – wet rot;
- Key issue 2 – Defect 2 – condensation;
- Key issue 3 – Remedial advice to the client.

The word count for the case study is 3,000 words, which is again absolute with no leeway. This includes everything between the end of the contents page and the start of the appendices. Candidates can be referred before being invited to interview by the RICS for exceeding the word count. As such, it should be treated as a strict client requirement. Candidates should avoid using the appendices to include additional detail which could not be fitted within the main body of the case study. This does not demonstrate adherence to a client's strict requirements and will not be viewed favourably by the assessment panel.

Rather than using footnotes, candidates should include core references to legislation or RICS guidance within the main body of the case study. This avoids the case study becoming overcomplicated and untidy.

The RICS provide a case study template to download from ARC, which should be followed rather than the candidate using their own format or structure. The following key sections should be included:

1. Introduction – Summary of the project, candidate's role, responsibilities, key stakeholders and timeline.
2. My Approach – This should discuss two to three key issues relating to the project and the challenges that the candidate needed to overcome. Some candidates may only use one key issue, but generally this should be avoided as the level of detail and advice is likely to be limited. Trying to include more than

three key issues is likely to lead to insufficient detail and analysis given the relatively limited overall word count. The candidate should explain the issue, options considered and the solution, advice and recommendations given. This requires analysis of the advantages and disadvantages of each option considered or of the specific considerations in giving reasoned advice to the client.

3 My Achievements – The candidate should discuss their key achievements, potentially with the use of subheadings for each competency demonstrated. This provides helpful structure and relates the actions and advice given back to the candidate's competency choices. The candidate should also discuss the outcome of the instruction and their overall achievements in terms of the level 3 reasoned advice they gave to their client.
4 Conclusion – The candidate should reflect on and critically analyse their involvement in their case study project, including lessons learnt and how they will improve on their performance in future. This section requires just as much thought as the preceding sections and demonstrates a candidate's commitment to their professional development. It is also advisable to discuss the candidate's professionalism and demonstration of ethics within the case study project, as this is a key aspect of being a Chartered Surveyor.
5 Appendices – The candidate should include an initial Appendix A to list the mandatory and technical competencies demonstrated within the case study. The other appendices should include only relevant illustrations, photographs or plans, which should be clearly referenced within the main body of the case study. Excessive or irrelevant appendices should not be included.

Confidentiality is a key ethical issue for candidates to consider. Candidates must have their employer's and client's express consent to disclose sensitive or confidential details within their submission. If this is not provided by either or both parties, then any specific details should be redacted, e.g. use Project X instead of the address, or secured lending client instead of the lender's actual identity. Anything that is litigious should also be redacted to ensure that identifiable details are not disclosed, for example the client's name, address or sensitive financial information. Candidates can be referred for not dealing with confidentiality appropriately, as it relates to a candidate's ethics and professionalism. The RICS also make a commitment to candidates that anything in the candidate's submission will not be disclosed further than the assessment panel.

What Does the CPD Record Include?

APC candidates must record at least 48 hours of CPD for every 12 month period. This means that candidates who need to record 24 months of structured training must record 24 months of CPD, with the minimum requirement recorded for each 12 month period. Candidates who need to record 12 months of structured training must record 12 months of CPD, with the minimum requirement recorded for this period.

All other candidates need to record their last 12 months of CPD activities. For preliminary review candidates, 12 months of CPD is required before both the preliminary review submission and the final assessment submission. The latter should be updated to reflect the relevant 12 month period.

Candidates do not need to submit CPD records for a period in excess of this, i.e., there is no requirement to submit CPD for a period longer than the required 12 or 24 months. In fact, submitting an extended CPD record could give rise to additional questioning that a candidate could otherwise avoid. Where this relates to historic CPD, the detail and outcomes may be difficult to remember or less relevant to the candidate's current role and experience.

A candidate would be much better served by including relevant CPD for the required period and in sufficient depth and detail. In particular, candidates should always remember that anything they write in their final assessment submission, including their CPD record, can be questioned by their assessment panel.

The CPD requirement is calculated by the RICS on a rolling basis dating back from the candidate's submission date. For example, if a candidate submits in February 2021 then their CPD must be undertaken in the 12 or 24 months, if appropriate, prior to this, i.e., from February 2020 (or 2019) to February 2021. The CPD requirements do not relate to the CPD recording requirements for qualified members, which are lower. This is a common source of confusion for APC candidates.

All CPD logged by APC candidates should relate clearly to the chosen technical and mandatory competencies, with relevance to the candidate's scope of work, role and responsibilities. Ideally, a candidate will undertake a variety of different types of CPD to maximise their learning opportunities. Candidates should also plan their CPD in advance and ensure that they reflect on what they have learnt afterwards, demonstrating their analytical and evaluative skills.

At least 50% of the minimum 48 hours' CPD must be formal in nature. This requires a structured approach with clear learning objectives. Examples of formal CPD (RICS, 2020i) include structured seminars, professional courses (which do not have to be run by the RICS), structured self-managed learning and online webinars.

CPD is logged by candidates via ARC, including the following key details:

- Description of the Activity – For example, 'Webinar – JCT Contracts' or 'Self-managed learning on the Red Book';
- Activity Status – Future or completed;
- Start Date;
- Time – Hours and minutes allocated to the CPD activity;
- Activity Type – Formal or informal and method, e.g., webinar, conference, private study or mentoring;
- Learning Outcomes – Reflection on what was learnt during the CPD activity. This should be written in the first person and past tense, e.g., 'I learnt about the different procurement methods and how they can be applied in my role', or 'I learnt about the new RICS Professional Statement and the core principles which I can apply to my involvement with commercial service charges'.

It is essential that candidates provide sufficient detail in their CPD record to allow the assessment panel to understand how the activity relates to a candidates' competencies, roles and experience. If this is poorly written or incomplete, then it may constitute grounds for referral. The same applies if insufficient hours have been undertaken, or if activities are informal rather than formal, meaning that the minimum requirements are not met. Candidates should also proofread their CPD record carefully and ensure that the detail provided is accurate, relevant and concise.

What Is the Ethics Module?

Candidates also need to complete the online RICS ethics module and test via ARC within 12 months of applying for final assessment. The module is also a helpful revision tool which candidates should consider using to prepare for their final assessment interview. Ethics is the only area for an automatic fail at the interview stage, so the importance and content of the ethics module should be held by candidates in high regard.

What Is the Preliminary Review Submission?

Candidates who do not have an RICS-accredited degree, i.e., a non-cognate degree, will need to undergo the additional preliminary review submission stage.

This essentially means that a candidate prepares their written preliminary review submission and submits this for a written review by the RICS, before they can submit their final assessment submission roughly four to six months later, or longer, if preferred.

The candidate's final assessment and preliminary review submissions are no different in terms of structure or format. It is simply a 'pre-check' process carried out by the RICS in the absence of an RICS-accredited degree, for example, which provides confidence that minimum standards of writing, structure and quality are being met.

The purpose of preliminary review is to determine if a candidate's submission, in terms of content, quality and professionalism, is of the standard required by the RICS. Essentially, 'Is the submission suitable for the APC assessors to prepare for and conduct the final assessment interview effectively?' (RICS, 2020j).

The preliminary review assessment panel will consider:

- Whether holistically the candidate's submission is of the required standard;
- Whether basic RICS requirements are met, including word count, professionalism, written English, structure and sufficient examples at levels 2 and 3;
- Whether content requirements are met, such as including relevant detail and evidence relating to a candidate's chosen competencies in their summary of experience and case study. This requires the candidate to reflect on the requirements of the pathway guide and competency descriptions within their written submission.

The results for preliminary review candidates typically take around eight weeks to be issued, on the last working day of the month. Substantial feedback will be provided within the feedback report whether or not the outcome is a pass or a referral.

If the submission is acceptable, then the candidate may submit their final assessment submission in a future submission window. Candidates and their Counsellors should closely scrutinise their feedback report to ensure they make any relevant improvements in their final submission.

Candidates may also wish to update their experience, examples and CPD activities to reflect new work undertaken following their preliminary review result being received. It is worth remembering that the preliminary review feedback is the subjective view of the specific panel at the time; sometimes it is not possible to make every single amendment particularly given the views or advice of a candidate's Counsellor or employer.

Candidates also need to ensure that their case study remains valid (i.e. within a two-year window) when they come forward for final assessment. This may not be the case if the preliminary review stage was passed more than six months prior, for example. In this case, the candidate may need to write a new case study, taking into account the general feedback provided at preliminary review. If a candidate does change their case study topic or examples included in their summary of experience, they do not need to submit for preliminary review again.

If the submission is not acceptable, then the candidate will receive a feedback report and will need to re-submit for preliminary review before they can proceed to submit for final assessment. Again, Candidates and their Counsellors should review the report together and ensure that improvements are made before the candidate resubmits for re-assessment at a future preliminary review window.

A successful preliminary review outcome is not a guarantee of success at the candidate's final assessment interview, which is based primarily on the candidate's performance at interview. It is also not an assessment of the candidate's competence, as the preliminary review assessors are only reviewing a snapshot written submission.

What Is the Assessment Resource Centre?

All candidates need to use the Assessment Resource Centre (ARC) to start, monitor and conclude their APC assessment journey. ARC is also used by the candidate's Counsellor for the process of supporting and signing off the final written submission.

ARC is an online RICS system which is used to upload the candidate's submission. Both the candidate and their Counsellor will have their own log-in and system access, with some of the information being shared between the parties, e.g. summary of experience and case study.

The candidate has their own home page on ARC which provides an overview of their progress. This includes the candidate's profile, candidate details, Counsellor details, progress overview and competencies. Changes to these can be made via the profile edit view, with the exception of the candidate's name, which needs to be notified to the RICS by email.

The candidate will need to contact the RICS to confirm any change in Counsellor details, as it is not possible to edit these directly via ARC. Once nominated, the candidate's nominated Counsellor will have access to the candidate's details via their own Counsellor ARC log-in.

The candidate's profile needs to be filled out as soon as a candidate is enrolled and has access to ARC. This requires a photograph of the candidate, contact details, pathway, qualifications (academic and professional), employment history and Counsellor number.

The candidate's final submission will include details of their qualifications and employment history, so it is important to ensure that these sections are filled out accurately and fully. Missing information will affect the assessors' ability to review the candidate's final submission. The candidate will need to search for their employer on the RICS database and if they are not listed, they will need to email the RICS to request they are added. The same may apply to academic or professional qualifications, which are not already listed on the system.

It is also very important to ensure that the candidate's photograph is professional and recent (within six months of applying for assessment), as it will be included on the front page of the final assessment submission. Rather than using a cropped photograph or one from a social occasion, candidates should try to take a simple, professional headshot using a mobile phone, tablet device, web camera or digital camera against a white or neutral background or wall.

Any change that a candidate needs to make to the information held on ARC requiring input or action by the RICS, e.g., Counsellor or employer additions will take time to be resolved. It is, therefore, highly recommended that any changes are made well in advance of the candidate's submission deadline. The Counsellor will also need to have undertaken the RICS Counsellor training before they are able to sign-off their first candidate.

The candidate will also need to select their competencies on ARC using the competency selector view. This includes confirming the mandatory and core competencies and selecting the relevant optional competencies. A candidate can edit their competency choices at any time before submitting their written submission, so these are not set in stone once confirmed on ARC. This may be relevant if a candidate's role or responsibilities change during the course of their APC journey or if they later feel that another choice would be more reflective of their day-to-day experience. Historic competencies are archived and remain accessible on ARC, in case a candidate makes further changes in future.

When preparing and uploading the written submission, candidates need to:

- Upload each separate summary of experience competency wording (for every level of every competency selected) to the individual boxes on ARC.
- Upload the case study in PDF format onto the case study view page. There is a Word Document template to download that candidates should use. Candidates will then need to convert this to PDF format (using Adobe Acrobat, the Word export function or an online convertor) prior to uploading to ARC. Any appendices, images or tables must be included within one single PDF, as multiple documents cannot not be uploaded. The case study can be uploaded multiple times, so if further changes are made, only the most recent version will be included within the final submission.
- Upload all CPD entries to the CPD record view. This requires each individual CPD activity to be uploaded with a description of the activity, activity status (completed/planned), start date, hours and minutes allocated, activity type (formal/informal) and learning outcome. Of particular importance is ensuring that the learning outcome is sufficiently detailed and written in plain, clear and accurate English. Ideally, the entries will have been uploaded over the course of the candidate's APC preparation, rather than at the last minute which can be an extremely time consuming task.
- Undertake the online ethics module and test within the twelve months prior to assessment.

Structured training candidates should also ensure that their diary is completed and meets the minimum RICS requirements, although this will not form part of the submission document. Candidates should record their diary entries as they go, to ensure that sufficient relevant detail is included. It can be a very lengthy task to record all the entries at the end of the APC process and this, in reality, defeats the objective of recording the diary entries.

Candidates should pay particular attention to the word counts for each section of their submission and ensure that they do not exceed these as there is no leeway for being over the word count. This is a referral point and can mean that a candidate is not able to proceed to the final online interview stage.

The Counsellor sign-off process can be challenging and requires time, so it is recommended that this is prepared for and undertaken well before the submission deadline.

To facilitate the feedback and review process by the candidate's Counsellor, it may be advisable to use a Word document rather than ARC as a working document, before the approved final competencies and case study are uploaded by the candidate to ARC. Using track changes and comments boxes can provide more in-depth feedback than the use of the feedback box on ARC, reducing the amount of time needed for the candidate and Counsellor to continually send back and forth feedback and amendments via ARC. This can be a frustrating process, particularly if the candidate uploads the wrong detail or needs to make minor amendments. The candidate will not be able to make these until the competency wording, for example, is returned to the candidate by the Counsellor via ARC.

The APC sign-off process via ARC requires the following actions:

- Candidate to upload their individual summary of experience entries (for every level of every competency) and case study via their own ARC system. They then need to click the request review button for their Counsellor to review and either approve or provide feedback on each element of their submission.
- When the individual elements of the submission have been approved, the Counsellor will need to approve the candidate for final assessment.

- When the Counsellor has approved the candidate for final assessment, the candidate will note that their ARC screen shows green completed boxes next to each element of their submission. This allows the candidate to apply for final assessment when the submission window opens. This is usually confirmed by a banner at the bottom of the candidate's ARC window confirming the relevant dates. Candidates cannot submit outside of the submission windows confirmed by the RICS.

The candidate's Counsellor should be notified in the messages view when the candidate's submission elements are ready for review. However, these are not followed up by email notification so the Counsellor and candidate should liaise closely to complete the sign-off process within the required timescales.

As part of the final submission process, candidates need to complete the proposer and seconder declaration process. This requires the candidate to provide contact details (RICS number or email address registered with RICS), for one proposer and two seconders. Either of these roles can be fulfilled by the candidate's Counsellor or Supervisor. All three parties need to be MRICS or FRICS members and no more than two of the parties can be from the candidate's employer. When the candidate has inputted the details of their proposer and seconder into ARC during the submission process, the three signatories will each receive an email requiring them to log into their own ARC account. There is then a simple approval process to follow, which will allow the candidate to proceed to final assessment. Without these signatories provided before the submission window closes, a candidate cannot proceed.

When the candidate is able to apply for final assessment, they have a variety of further choices and options to select via ARC:

- International Experience – If relevant, the candidate should confirm this here;
- Special Considerations – Including any extenuating circumstances or conditions the assessors should be aware of, e.g., dyslexia;
- Specialist Area – Some pathways and RICS regions require candidates to select a specialist area. In the UK, selection of a specialist area is required for the following pathways: Commercial Real Estate, Residential, Valuation, Quantity Surveying and Construction, Infrastructure, Geomatics, Minerals and Waste

Management, Planning and Development and Land and Resources. The reason for the specialist area selection is to enable the RICS to appoint assessors who will be sufficiently experienced and knowledgeable in the candidate's area of practice and chosen competencies;

- Interview Dates – Candidates can select priority weeks for their potential assessment date. However, this will only be confirmed by email three weeks beforehand. Candidates should, therefore, keep themselves free during the weeks selected during the submission process;
- Download Submission – The candidate should download their PDF submission to ensure that they are happy with the final document. ARC may make formatting adjustments in the spacing based on the candidate's work, which may affect the presentation of the final submission. Unfortunately, there is no way to amend this and the assessment panel will be aware of this potential issue. The candidate will, therefore, not be penalised for any formatting or spacing issues due to ARC in their written submission.

The candidate will then need to confirm they are ready to submit for assessment, and, when complete, this will be confirmed by a successful confirmation banner at the bottom of the page on ARC. There is no leeway for late submissions, i.e., ARC will not allow the candidate to submit late, so diligent forward planning should be undertaken by candidates. If the submission window is missed, then the candidate will need to wait until the next submission window opens.

Candidates applying for preliminary review assessment have a very similar process to follow via ARC. One of the key differences is that the candidate does not need to complete the proposer and seconder declaration process as this only applies for the final assessment submission.

What Is the Final Assessment Interview?

After candidates have submitted their final assessment submission via ARC, they will be notified of their final assessment interview date and time three weeks beforehand by email. All interviews are online and no further assessments are planned by the RICS to take place face-to-face.

If a candidate submits their final assessment documents but then subsequently wishes to defer until the next sitting, they can do so by emailing the RICS. No charge is payable if notification of the deferral is provided within fourteen days of submitting, however, a charge will apply thereafter. Candidates may decide to defer for a variety of reasons, e.g., extenuating circumstances affect their availability or if they feel that they are not ready to be assessed.

The assessment panel will consider the following during the candidate's final assessment interview:

- The candidate's communication skills both through a ten minute presentation on their case study and responses to the assessors' questioning on their final assessment submission;
- How well the candidate is able to verbalise and explain their advice and actions discussed within their written submission, demonstrating the stated competency levels;
- That the candidate understands the role and responsibilities of a Chartered Surveyor, including advising clients diligently and acting ethically and professionally;
- That the candidate is able to act independently and unsupervised, i.e., that they are a safe pair of hands. This is particularly relevant because after qualifying as MRICS, a candidate could in theory set up in practice as a sole trader. The assessment panel need to be confident that any candidates who are successful in their APC would be competent to do this ethically and professionally.

The assessors will assess the candidate's written submission, although this will be considered holistically within the context of the interview. Unless there is a clear deficiency or issue in terms of word count or basic requirements of the written submission (e.g., CPD hours), then a candidate will not be referred on their written submission alone. Any substantial issues, such as the word count being exceeded, should be picked up by the RICS after the candidate submits via ARC and the candidate would then not be permitted to proceed to the final interview stage.

Candidates will be assessed by a panel of two or three Chartered Surveyors, one chairperson and either one or two assessors. All will be trained APC assessors or chairs and experienced in providing final assessment interviews. To ensure that the interview is tailored to the candidate's experience, two of the panel members will be from the same pathway as the candidate. Some pathways also have specialist areas and at least one member of the panel will be experienced in this area.

Assessors are trained to give each and every candidate a fair interview process and to treat all candidates with respect and dignity. The interview process should encourage and facilitate all candidates to succeed if they meet the required levels of competence. In fact, the online assessment process means that an even more diverse range of assessors are available to participate in the assessment process and this is likely to give candidates a much more positive, fair and transparent interview experience.

Candidates should be made to feel at ease by their panel, who should be positive and provide candidates with an environment in which they can excel. Ways they may do this include:

- Asking open questions, one at a time, to avoid being confusing or overly elaborate;
- Being flexible in their questioning approach and following up on a candidate's answers with further questions if appropriate;
- Supporting a candidate and encouraging them if they are nervous or stressed.

However, panels will not give candidates any indication of how they are doing during the interview. They will not make any recommendations or suggestions as to whether the candidate has passed or failed. They will not use any affirmations or provide feedback to the candidate during the interview.

The best indication of a correct answer, therefore, is sometimes the panel moving onto another competency or question. If an assessor is pursuing the same line of questioning or driving at a specific point, this may be an indicator that the candidate's answers need further thought or clarity. At this point, asking for clarification, or returning to the question at the end of the interview may be sensible.

A staff facilitator may also be present during the interview to assist the candidate and assessment panel with any technical issues. An auditor may also be present to observe the assessment panel's performance, rather than that of the candidate.

The assessment panel should have no perceived or actual conflicts of interest in assessing the candidate fairly and transparently. This could be either personal, e.g., a candidate and assessor having met at a CPD event or being familiar in a professional context, or prejudicial, e.g., where an assessor may benefit from the candidate's success. A personal conflict may not present an issue if both parties are happy to proceed, however, a prejudicial conflict should be declared beforehand and the assessor removed from the panel. If a candidate becomes aware of a conflict of interest on the day, then they should make their chairperson aware. The matter will be dealt with either by a two-person panel proceeding or the interview being rescheduled with a non-conflicted panel.

Candidates should have made the RICS aware of any special considerations or extenuating circumstances during the ARC submission process. The RICS may require supporting medical evidence to be provided to enable relevant reasonable adjustments to be made. If a candidate feels that the panel may benefit from being made aware of how any specific issues affect them, they may wish to prepare a written statement to read out to the panel during the initial welcome at the start of the interview.

The interview lasts for sixty minutes and is conducted online, currently using Microsoft Teams. Candidates should ensure that they test out the system beforehand and have a good quality video camera and microphone. There is no excuse for hardware or systems not working on the day.

Candidates can prepare effectively for the interview by choosing a quiet interview location, where disturbances are kept to a minimum. This may mean asking others to keep quiet or to leave the location for the duration of the interview. If a candidate's home environment is too noisy, then they may prefer to sit their interview in an office or external meeting room.

Candidates should also ensure they have good quality Wi-Fi or tether to 4G if this is not available. This should be tested beforehand and if weak, a candidate should reconsider their choice of location or ask other users to switch their devices off during the interview. Any devices used should be connected to a power source rather than relying on battery power, which is quickly drained.

Another important factor to consider is having good lighting and a background that is neutral and clear. Candidates could sit near to natural light or have a lamp next to or behind them. They should also avoid sitting in front of a window which can be blinding to the camera. The camera should be at eye height to create a natural connection with the assessment panel.

Candidates should ensure that their interview environment is comfortable, including having a supportive chair, adequate height desk, notepad, pen and glass of water. Candidates may also choose to have a second screen, such as a tablet or monitor, to use for their presentation notes only. To ensure good quality audio, candidates should consider using headphones or a separate microphone.

Smartphones should be avoided, as they are typically too small to allow an effective interface during the interview. Candidates should ensure they close any applications not being used to minimise the risk of technical issues, as well as putting any devices on silent or do not disturb.

Candidates should wear smart, professional clothing appropriate for a client meeting or job interview. Ideally this should be simple and plain to avoid being distracting on camera.

Candidates are not permitted to record their interview and any attempt to do so may lead to disciplinary action and immediate termination of the interview.

The sixty minute time limit is strict and will be managed closely by the chairperson. In the event of technical difficulties, additional time up to ten minutes will be added to the end of the interview. If interruptions total more than ten minutes then the chairperson will terminate the interview and it will be rescheduled.

A candidate's interview may also be extended if any special considerations or extenuating circumstances dictate that additional time should be given to allow the candidate to respond fully to questions asked. In this case, the candidate will not be asked more questions than they would otherwise be in a standard sixty minute interview. It is, instead, the amount of time that they are given to listen, comprehend and respond to questions that may be extended.

Candidates should join their interview video link five to ten minutes beforehand. They will then be permitted access from the virtual lobby when their assessment panel is ready.

The interview will be structured strictly as follows:

- Initial welcome by the chairperson before the sixty minute time limit starts. This will include the candidate being asked to complete a 360 degree view of their surroundings to ensure that no outside assistance is being provided. The chairperson is at liberty

to ask for this to be repeated at any time during the interview. The chairperson will then explain the interview structure and ask if the candidate is fit, well and ready to proceed;
- Ten minute presentation by the candidate, focussing on their case study;
- Ten minutes of questioning by the assessors on the candidate's case study;
- Thirty minutes of questioning and discussion by the assessors on the candidate's summary of experience, CPD, Rules of Conduct and professional practice;
- Ten minutes' questioning by the chairperson on any outstanding competencies and ethics. This will include the chairperson's closing comments and the opportunity for the candidate to have the last word. This may include coming back to any questions the candidate was unable to answer earlier on or to make any additional comments, which can be noted down on a piece of paper during the interview.

The candidate's case study presentation could focus on one or more of the key issues. However, it should not simply repeat verbatim what a candidate has written. The presentation could be made interesting by delving deeper into one of the key issues, looking at progress of the project since the case study was written or discussing further the candidate's analysis of the options and advice given to the client. The main aim of the presentation is to demonstrate strong communication and presentation skills to the panel. Therefore, the presentation does not need to be complicated and it does not necessarily need to introduce anything new to the panel.

The presentation can be supported by a visual aid. This can either be screen shared with the panel or physically shown to the camera. Equally, a candidate may choose not to use a visual aid if they feel that it will not add to their presentation.

The best visual aids are generally simple and used only to support key points within the presentation. This could be via a brief number of clear PDFs screen shared at optimal points during the presentation. Using a full slideshow is not recommended as the candidate will not be seen clearly on video and it can detract from the presentation. It is also not recommended to use a flipchart or visual aid physically held up to the camera, as this is likely to be hard to read by the assessment panel. Visual aids should be professionally presented, with a clear title and large, easy to read text or graphics. Given the reliance on IT technology to screen share, candidates should be prepared not to use a visual aid in the event of technical difficulties.

Candidates can use cue cards or brief notes for their presentation. During the rest of the interview, they will not be able to use any notes or have a copy submission to hand.

The timing of the case study presentation is key and candidates will be asked to stop if they exceed this by more than ten to fifteen seconds. Equally, being substantially below this will be a negative consideration in the context of the assessors' overall decision. Candidates should practise their presentation frequently to ensure that they are fluent in giving it and are able to meet the timing requirement accurately. A stopwatch, clock or timer can be used, providing that it is not overly distracting to the candidate and their assessment panel.

Given that the interview is conducted online, it is vitally important for candidates to make the best use of both verbal and non-verbal communication. Candidates should aim to speak clearly and concisely, taking time to understand and listen to questions before answering. They should also be aware of avoiding too much movement or gesticulation, which can be distracting on camera. Positive body language will help to portray a confident candidate and practising this on camera beforehand with a friend, family member or colleague can be helpful. At all times, candidates should try to look directly into the camera and maintain good eye contact with their assessment panel.

The interview is based on the candidate's written submission. This means that anything included in the candidate's case study, summary of experience and CPD record can be questioned by the assessment panel. Therefore, candidates may wish to limit references to complex caselaw, for example, in their written submission, unless they are very confident to discuss these in their interview. Candidates should also be aware that they will be assessed in relation to the legislation and guidance relating to the geographic region of their final assessment. This means that if experience is declared from other countries, candidates will need to have knowledge of legislative and regulatory requirements for both regions.

Candidates may also be asked on current hot topics regarding industry or market issues, providing they are relevant to the candidate's area of practice. This means that having a good level of market awareness is important, which can be obtained through reading trade press and a good quality newspaper, listening to relevant podcasts and watching CPD videos.

Assessors are trained to begin questioning at the highest level declared, with supporting level 1 knowledge-based questions potentially being asked in order to explore the justification for the advice or actions of the candidate. Candidates should ensure they are familiar with any examples included in levels 2 and 3, as these should form the basis of the majority of their answers. Candidates will not be asked questions on competencies they have not selected or at levels beyond those declared, i.e., they will not be expected to give reasoned advice (level 3) if they have declared a competency only to level 2 (acting or doing).

Questions posed by the panel should not be hypothetical and they should encourage the candidate to answer based on their experience. All competencies will be questioned by the assessment panel, with ethics questions included within the main body of the interview, if possible. Additional ethics questions will also be asked in the final ten minutes by the chairperson. Candidates should be aware that elements of their mandatory competencies may be assessed within their technical competency questioning. A candidate's communication skills will also be assessed during their case study presentation and response style to the panel's questions.

Assessors are trained to signpost candidates to the competency area being questioned. Candidates should ask for clarification if they are unsure of the question posed to them or if they are not sure of the area of focus being sought by the assessor. Candidates should also ensure they listen carefully to the questions they are being asked, take a deep breath before answering and then seek to give only the response required by the assessors.

The overall assessment decision is holistic and one wrong answer will not constitute a referral. This means that poor performance in one competency may be balanced out by good performance elsewhere. However, an inability to demonstrate a number of competencies to the required level is likely to constitute a referral, particularly if these are the competencies declared by the candidate at level 3. Candidates are not expected to be experts in every area of their professional practice. The assessors are, therefore, seeking to confirm that the candidate has met the minimum required levels of competence declared.

Candidates should also be mindful that acting ethically is at the heart of what it means to be a Chartered Surveyor. Ethics, Rules of Conduct and Professionalism is, therefore, the only competency where a wrong, or unethical answer, will constitute an automatic fail.

The assessment is not designed to be an exam. It is an assessment of professional competence and there will be questions that a candidate cannot answer. This requires a toolbox of potential responses to be practised by candidates. For example, some questions may relate to experience or knowledge outside the core scope of competence or practice of the candidate. For these questions, the candidate can state this and identify where they would seek specialist advice, input or support from. This shows that the candidate is able to take responsibility and is a safe pair of hands when dealing with clients.

Candidates may also encounter questions they simply cannot answer due to the pressure of the interview process. In these situations, being able to refer or signpost the assessors to suitable RICS guidance or documents is helpful. When under pressure, candidates should avoid trying to explain to the assessors everything they know about a topic or issue. This does not show diligence or a considered approach to advising clients. Instead, a candidate would be better placed to come back to the question later or use some of the tools and tactics outlined above.

The final result will be decided by the assessment panel straight after the interview ends. In the case of a three person panel, the assessors will make a decision and if a split decision is reached, the chairperson will make the final decision. The same applies for a two person panel, where the chair and assessor will make their own decisions, although the chairperson again has the final say.

What Does the Senior Professional Submission and Assessment Include?

The senior professional assessment is one of three separate APC assessment routes. This requires candidates to have at least ten years of relevant experience, which is reduced to five years if a postgraduate degree is held. Candidates also need to be able to demonstrate senior professional responsibilities in their role, including leadership, management of people and management of resources.

The RICS define a senior professional as 'an individual with advanced responsibilities who is recognised for their impact and career progression within the profession' (RICS, 2020k). This requires demonstrating the qualities (or indicators) of leadership, managing people and managing resources.

The senior professional assessment is, therefore, appropriate for candidates in senior management positions or who are responsible for managing or leading teams, rather than being responsible for day-to-day surveying work. Indicators of this assessment route being appropriate are a candidate's position in their organisation's structure, having decision-making responsibilities, an international dimension to their role, a high profile client base and recognition from the wider industry, peers and media.

Senior professionals should also be able to demonstrate at least one of the following behaviours in their work:

- Pursuing opportunities to develop the industry and the profession;
- Advocating best practice;
- Taking responsibility to deliver professionalism;
- Acting with integrity to promote responsible business.

Senior professional candidates need to satisfy each of three senior professional competencies to level 2, as well as the mandatory and technical (core and optional) competencies relevant to their chosen pathway.

The senior professional competencies are:

- Leadership;
- Managing people;
- Managing resources.

Senior professionals must undergo an initial vetting stage by submitting an application form to the RICS including their employment history, qualifications, pathway and 400-word senior profile statement.

The 400-word senior profile statement should discuss the candidate's role, activities and senior professional responsibilities, referencing the indicators and behaviours detailed above. The candidate should also include relevant detail on their activities and impact as a senior professional, with a supporting organogram showing their senior professional position in their organisation.

Successful candidates are eligible to enrol for the senior professional assessment and must submit their final assessment submission within twelve months. If this deadline is missed, then the candidate will need to re-apply for the initial vetting stage by the RICS.

The senior professional written final assessment submission includes the following elements:

- Application form, as submitted at the initial vetting stage;
- CPD record, although the requirements are different to the 'traditional' APC routes. Candidates require twenty hours' CPD for the last twelve months prior to submitting, including at least 50% comprising formal activities;
- RICS ethics module and test certificate, which needs to be dated in the twelve months prior to submitting for final assessment;
- Competency and pathway selection;
- Three case studies of 1,000 to 1,500 words each.

The three case studies each focus on a single project or instruction, where the candidate has demonstrated their senior professional role in terms of management, leadership, client relationship management or strategic advice. The candidate may have delegated or overseen provision of some of the technical input by employees, consultants or contractors.

The three case studies have different requirements:

- Senior professional case study – Focussing on a project demonstrating the senior professional competencies;
- Technical case studies 1 and 2 – Focussing on different projects in each case study that demonstrate the candidate's experience against at least two core technical competencies, of which at least one must be demonstrated to level 3, i.e., where they have given reasoned advice. These case studies should also demonstrate the three senior professional competencies, in addition to any relevant mandatory competencies and the candidate's appreciation of ethics and professionalism.

When selecting appropriate case study projects, candidates should bear in mind the following:

- The case study must have been undertaken within the last three years, calculated on a rolling basis from the submission date. Some projects will have extended over a longer period of time, in which case the candidate's primary involvement must have been in the last three years;
- The candidate should include at least one case study which has been undertaken in the country they will be assessed in, allowing them to demonstrate relevant knowledge of national legislation and guidance;
- The project could have been provided to an internal or external client;
- Candidates need to have express client and employer's consent to use the chosen case studies or, if this is not possible, to redact any details which make the project identifiable.

Each case study should follow a consistent structure, including the following elements:

- Overview of the project, objectives and key issues;
- Role in the project, including responsibilities, actions and reasoned advice;
- Analysis of key issues, challenges or problems and the solutions or remedial action recommended. This will focus on the candidate's approach and delivery of key objectives or outcomes;
- Conclusion looking at the key achievements and wider impacts on the client, employer, candidate's career and future work;
- Statement of specific competencies which have been demonstrated;
- Appendices, such as drawings, photographs or plans. These should not be extensive and kept to only those which are directly relevant to the case study content.

The case studies should be carefully proofread to demonstrate high standards of professionalism.

After submitting the final assessment submission, candidates will be invited to a sixty minute online final assessment interview by the RICS. This aims to assess whether the candidate has:

- Applied their level 1 knowledge through professional experience and advice (levels 2 and 3) and recognised the impact of their senior professional role and responsibilities on the relevant sector or market. This requires the candidate to demonstrate clear comprehension of their selected competencies and pathway;

- Acted ethically and in accordance with their duty of care to clients, employers and the wider public;
- Acted as an ambassador for the profession;
- Acted in pursuance of the objectives of their client or employer;
- Demonstrated understanding of up-to-date and geographically relevant legislation and technical theory.

The sixty minute interview is structured as follows:

- Ten minutes – Candidate's presentation providing a personal introduction and background on their senior professional career history and current role. This should not focus specifically on the case studies, as is required for the 'traditional' APC assessment;
- Fifty minutes – Questioning and discussion of the candidate's submission, senior professional role, responsibilities and wider professional issues, including ethics.

The weighting of the panel's decision will be allocated as 50% towards their senior professional profile, 25% towards the pathway competencies and 25% towards ethics and professionalism.

The interview is not an exam and the assessment panel will structure it as a professional discussion, with a focus on exploring the candidate's senior professional role and background. The questioning will be based on the submitted documents, but with wider issues investigated to ensure that the candidate has reached the level required of a Chartered Surveyor.

There will be an appreciation and understanding by the assessment panel that senior professional candidates will often be managing or delegating work to others, rather than undertaking day-to-day technical work themselves. This means that the emphasis will be on the candidate's management and leadership skills, alongside a strong focus on ethics and professional behaviour, particularly when leading or managing other professionals, contractors or consultants.

What Does the Specialist Submission and Assessment Include?

Similarly to the senior professional assessment, the specialist assessment requires candidates to have at least ten years of relevant experience, which is reduced to five years if a relevant undergraduate degree or relevant professional qualification and a relevant postgraduate degree (master's or higher) are held.

Specialist candidates need to have advanced responsibilities in a specialist or niche area of work. The RICS define a specialist as 'an individual delivering enhanced services who is recognised for their impact and authority within the profession' (RICS, 2020l).

Being a specialist may also include the following indicators:

- Having a decision-making position;
- A track record of specialist consultancy work;
- Speaking at conferences;
- Writing in the trade press;
- Being appointed by a governance/judicial body;
- Being recognised by the wider industry, peers and media;
- Lecturing;
- Providing formal training;
- Being qualified above master's level (e.g., PhD);
- Being involved in dispute resolution for a technical area.

Specialists should also be able to demonstrate one or more of the following behaviours in their work:

- Pursuing opportunities to develop the industry and the profession;
- Advocating best practice;
- Taking responsibility to deliver professionalism;
- Acting with integrity to promote responsible business.

The initial stage of applying for the specialist assessment involves undergoing the vetting stage. This requires completion of the specialist application form, including template CV, chosen APC pathway and 400-word specialist profile statement.

When choosing a relevant pathway, specialist candidates should ensure they are able to satisfy one or two core technical competencies relating to their specialist area of work. At least one of these must be declared at level 3.

The specialist CV requires the candidate to confirm their academic education, professional qualifications and professional experience (including an overview of the scope and responsibilities of each role). It also requires the candidate to confirm the specialist indicators demonstrated in their work from the following:

- Position in the organisation structure;
- Publications;
- Record of specialist consultancy work;
- Record of speaking at high level conferences;
- Dispute resolution;
- Recognition;
- Appointment by governance or judicial body;
- Record of lecturing or formal training;
- Qualifications.

The specialist profile statement of 400 words should provide a clear overview of why the candidate is eligible for the specialist assessment. The candidate should discuss their specialist services and activities, together with the behaviours that demonstrate the authority and impact of their specialist role. The statement should include an organogram or description of the specialist's role in their organisation's structure.

The vetting form will be reviewed by the RICS against the following criteria:

- Sufficient experience;
- Relevant qualifications;
- Experience in the chosen pathway;
- Reference made to the required specialist indicators and at least one specialist behaviour;
- Evidence of specialist authority, enhanced services and outcomes/impact.

Successful candidates are eligible to enrol for the specialist assessment and must submit their final assessment submission within twelve months. If this deadline is missed, then the candidate will need to re-apply for the initial vetting stage by the RICS.

The specialist written final assessment submission includes the following elements:

- Application form, as submitted at the initial vetting stage;
- CPD record, although the requirements differ from the 'traditional' APC routes. Candidates require twenty hours' CPD for the last twelve months prior to submitting, including at least 50% comprising formal activities;
- RICS ethics module and test certificate, which needs to be dated in the twelve months prior to submitting for final assessment;
- Competency and pathway selection;
- Three case studies.

The three case studies have different requirements, although all need to relate to the candidate's specialist profile:

- Specialist case study – Focussing on a project demonstrating specialist experience in one or two of the candidate's chosen technical core competencies, of which at least one must be to level 3 (i.e. giving reasoned advice);
- Technical case studies 1 and 2 – Focussing on different projects in each case study that demonstrates the candidate's experience against at least two technical competencies.

Each case study has a word count of 1,000 to 1,500 and should focus on different technical case studies. The case studies should also make relevant references to the mandatory competencies, in particular Ethics, Rules of Conduct and Professionalism, where appropriate.

When selecting appropriate case study projects, candidates should bear in mind the following requirements and considerations.

The case study must have been undertaken within the last three years, calculated on a rolling basis from the submission date. Some projects will have extended over a longer period of time, in which case the candidate's primary involvement must have been in the last three years.

The candidate should include at least one case study which has been undertaken in the country they will be assessed in, allowing them to demonstrate relevant knowledge of national legislation and guidance.

The project could have been provided to an internal or external client. Furthermore, the candidate should have led the project and been involved in setting strategy, decision-making, analysing options, recommending solutions and managing the client relationship. However, some of the 'day-to-day' technical work may have been delegated to employees, consultants or contractors.

Candidates need to have express client and employer's consent to use the chosen case studies or, if this is not possible, to redact any details which make the project identifiable.

Each case study should follow a consistent structure, including:

- An overview of the project, objectives and key issues;
- The role of the specialist in the project, including responsibilities, actions and reasoned advice;
- An analysis of key issues, challenges or problems faced and the solutions or remedial action recommended. This will focus on the candidate's approach and delivery of key objectives or outcomes;
- A conclusion looking at the key achievements and wider impacts on the client, employer, candidate's career and future work;
- A statement of specific competencies which have been demonstrated;
- Appendices, such as drawings, photographs or plans. These should not be extensive and kept to only those which are directly relevant to the case study content.

The case studies should be carefully proofread to demonstrate high standards of professionalism.

After submitting the final assessment submission, candidates will be invited to a sixty minute online final assessment interview by the RICS. This aims to assess whether the candidate has:

- Applied their level 1 knowledge through professional experience and advice (levels 2 and 3) and recognised the impact of their work on the relevant sector or market. This requires the candidate to have understood and to demonstrate clear comprehension of their selected competencies and pathway;
- Acted ethically and in accordance with their duty of care to clients, employers and the wider public;
- Acted as an ambassador for the profession;

- Acted in pursuance of the objectives of their client or employer;
- Demonstrated understanding of up-to-date and geographically relevant legislation and technical theory.

The sixty-minute interview is structured as follows:

- Ten minutes – Candidate's presentation providing a personal introduction and background on their specialist work and case studies;
- Fifty minutes – Questioning and discussion of the candidate's submission, specialist area of work and wider professional issues, including ethics.

The weighting of the panel's decision will be 50% towards their specialist profile, 25% towards the pathway competencies, and 25% towards ethics and professionalism.

The interview is not an exam and the assessment panel will structure it as a professional discussion, with a focus on exploring the candidate's specialist role and background. The questioning will be based on the submitted documents, but with wider issues investigated to ensure that the candidate has reached the level required of a Chartered Surveyor.

What Does the Academic Submission and Assessment Include?

The academic assessment is appropriate for academic professionals, e.g., lecturers or researchers. The assessment requirements reflect the differences in how competence will be demonstrated by academics as opposed to practising surveyors.

Eligible academics must have at least three years of academic experience and a surveying-related degree. The academic experience does not have to be continuous and can have been undertaken at various points in time.

Academic candidates must satisfy the same mandatory competencies as for all other APC candidates, irrespective of route or pathway. Academics must also select a pathway aligned to their area of academic practice, e.g., Commercial Real Estate or Quantity Surveying and Construction. This will inform the candidate's technical competency choices, where at least one level 3 core competency from the chosen pathway must be selected, alongside other core competencies applicable to all academic candidates. These include Data Management to level 2 and either Research Methodologies and Techniques or Leadership to level 3. The selection between the last two choices will depend on if the candidate has leadership and management experience (i.e., Leadership to level 3) or if their role is more focussed on data collection and analysis (i.e., Research Methodologies and Techniques or Leadership to level 3).

Academic candidates must initially submit their CV to the RICS using the academic CV template available on the RICS website. This includes the following:

- Personal details;
- Business details;
- Pathway selection;
- Education history;
- Professional qualifications;
- Professional experience for at least the last three years, including employers, roles and an overview of the candidate's scope and responsibilities;
- Academic criteria;
- Academic review statement, relating to the four pieces of evidence selected above.

The academic review statement comprises a 3,000 word summary of the candidate's academic experience and how four pieces of relevant evidence relate to their chosen pathway and the wider surveying profession. The four pieces of evidence do not need to be submitted to the RICS at this time. These only need to be submitted with the candidate's final assessment submission later in the process.

Competence needs to be demonstrated within three main areas, with candidates submitting four pieces of evidence in total, from the below three lists, to support their application:

- Teaching – Preparation and delivery of learning material, assessment of undergraduate and postgraduate student work through summative and formative marking and feedback, completion of a post-graduate teaching qualification, fellowship of the Higher Education Academy, mentoring and supervision of research students and course leadership and development;

- Research and Scholarship – Publishing of built environment or surveying-related research in journals, conferences proceedings, consultancy reports, government research, legal reports and books. Candidates should ensure that clear references are included for any publications mentioned, which should be recent, relevant and of national significance or quality;
- External Engagement and Academic Activities – Including embedding research, employability or professional practice into undergraduate or postgraduate level curricula, industry engagement and knowledge transfer. This could include student liaison groups, guest lectures, CPD events or engaging with an RICS Committee or Board.

At least one piece of evidence must evidence level 3 in one of the candidate's core competencies, as discussed above. Furthermore, at least one piece of evidence needs to come from each of the categories below, with the fourth piece of evidence coming from any of the categories depending on the candidate's area of academic focus. Candidates must have express consent to include the evidence in their submission from their employer and clients.

The academic review statement should be structured as a professional report, using sub-headings reflecting the following recommended sections:

- Introduction – Approximately 100 words explaining the candidate's career history, academic profile and role;
- Teaching (Evidence 1) – Approximately 700 words explaining how the candidate's teaching experience is relevant to their pathway and the surveying profession;
- Research and Scholarship (Evidence 2) – Approximately 700 words explaining how the candidate's research and scholarship experience are relevant to their pathway and the surveying profession;
- External Engagement and Academic Activities (Evidence 3) – Approximately 700 words explaining how the candidate's external engagement and academic activities are relevant to their pathway and the surveying profession;
- Additional Item (Evidence 4) – approximately 700 words explaining how the candidate's final piece of evidence is relevant to their pathway and the surveying profession.

The CV and academic review statement will first be reviewed by the RICS to ensure that the academic pathway is appropriate for the candidate. If approved, these two documents will then be further reviewed by an academic review panel who will confirm if the candidate can proceed to final assessment on the basis of the quality and content of the documents submitted. The academic review panel will consider the candidate's eligibility based on their academic profile, rather than assessing their professional competence which will be assessed in the final assessment interview.

The academic review panel report and feedback will be issued within twenty-one days. If the candidate is not successful, then the RICS will suggest alternative routes that would be a better fit for the candidate's experience and role.

If a candidate is successful at the second stage review, then they will be able to submit for final assessment and interview.

The final assessment submission, based on templates downloaded from ARC or the RICS website, should include the following:

- CV, as submitted at the initial review stage;
- Academic review statement, as submitted at the initial review stage;
- Four pieces of supporting evidence, which will not have been submitted at the initial review stage, although they will have been discussed in the candidate's academic review statement. The evidence should relate closely to the candidate's mandatory and technical (core and option) competencies.
- Summary of experience;
- CPD record – The requirements are the same as for the 'traditional' APC routes, i.e., forty-eight hours for the last twelve months prior to submitting, unlike for the senior professional and specialist assessments;
- RICS ethics module and test certificate, which needs to be dated in the twelve months prior to submitting for final assessment.

For the summary of experience, candidates should refer to the detail above relating to the 'traditional' APC routes. The requirements are very similar, although with a focus on the candidate's academic role and experience for the academic pathway. In particular, candidates should ensure that at level 3 they provide sufficient detail and practical experience relating to their academic practice at higher education level. The word counts are 1,500 for the mandatory competencies and 3,000–4,000 for the technical (core and optional) competencies. This roughly equates to 150–200 words per level and per competency.

Candidates who submit for final assessment will subsequently be invited for an online interview by the RICS. This lasts for sixty minutes with a panel of two or three assessors, at least one of whom will specialise in the academic route.

The interview will be structured as follows:

- Five minutes – Chairperson's opening and introduction;
- Ten minutes – Candidate's presentation on their career, role and one of the four pieces of evidence, specifically highlighting the candidate's academic skills. A visual aid can be used; candidates should refer to the interview advice earlier on in this chapter;
- Fifteen minutes – Questioning on the candidate's presentation;
- Fifteen minutes – Questioning and discussion on the candidate's wider submission and academic role;
- Ten minutes – Questioning and discussion on CPD, Rules of Conduct and professional practice;
- Five minutes – Chairperson's close and opportunity for the candidate to make any final comments or requests for clarification.

The panel will specifically be assessing the candidate in relation to the following criteria:

- Breadth of relevant academic experience and knowledge and transference of this to students, researchers or external consultants;
- Requirements of the chosen mandatory and technical (core and optional) competencies and wider pathway context;
- Oral and written communication skills through the written submission, presentation and interview responses and discussion;
- Understanding of the role and responsibilities of a Chartered Surveyor;
- Ethical and professional attitude with a clear duty of care provided to clients, employers and wider stakeholders;
- Being a good ambassador for the profession.

How Do Candidates Receive their Results?

All candidates will receive notification of their assessment result within five working days of their final assessment interview via ARC and shortly after via email.

For senior professional, specialist and academic candidates, the result may take longer to come through, generally seven days for senior professionals and specialists and twenty-one days for academics.

After qualifying, candidates will be added to the RICS Global Members Directory within twenty-four hours, a public announcement will be placed on the RICS website a few days later and an Award Pack issued by post six to eight weeks later. Candidates will also need to pay an upgrade fee to reflect their MRICS qualification status. Concessions may apply to certain professionals and the RICS can advise on eligibility, for example, academics working in academia rather than in a surveying role.

Conclusion

After reading this chapter, candidates should feel confident to start or continue their APC journey. Further chapters provide information on the appeals and referral processes. An additional fee is payable to the RICS in both instances.

5 AssocRICS

Introduction

In this chapter, we look at the process, structure and individual elements of the AssocRICS qualification. AssocRICS, unlike the APC, is based on a written submission only. There is no face-to-face interview or further assessment after the written submission has been passed by the RICS.

What Is AssocRICS?

AssocRICS is the first level of RICS qualification and a major achievement in its own right.

AssocRICS status demonstrates to a candidate's clients, peers and the public that they are competent to undertake their role. It does not provide Chartered Surveyor status, but can provide a steppingstone to MRICS, via the APC, if a candidate wishes to follow this route.

Many Residential pathway candidates, for example, decide to pursue AssocRICS only as it provides the ability to become an AssocRICS Registered Valuer. This may allow a surveyor to undertake secured lending or level 2 HomeBuyer Reports, for example.

AssocRICS is also often pursued by candidates who work in supporting roles within the industry, such as property specialists in an organisation's finance department or an asset manager with a wider role than just surveying.

What Are the Eligibility Requirements for AssocRICS?

AssocRICS is available to candidates who have the following:

DOI: 10.1201/9781003156673-5

- One year's relevant experience and a relevant undergraduate degree;
- Two years' relevant experience and a relevant higher, advanced or foundation level qualification;
- Four years' relevant experience and no academic qualifications.

Candidates have six years to pass the AssocRICS assessment from when they enrol online, similarly to the APC. The RICS set the 'six year rule' to ensure that candidates are committed towards achieving the qualification. After passing AssocRICS, candidates who wish to pursue the APC are no longer restricted by this rule and can take as long as they need to pursue MRICS status.

Candidates will initially need to enrol on the AssocRICS assessment via the RICS website. They will then be provided with access to the ARC system, which is where the final assessment submission is uploaded to.

What Are the AssocRICS Pathways?

AssocRICS candidates are able to choose from fourteen sector pathways:

1. Building Control;
2. Building Surveying;
3. Commercial Real Estate – Commercial Property Management;
4. Commercial Real Estate – Real Estate Agency;
5. Facilities Management;
6. Geomatics;
7. Infrastructure;
8. Land and Resources;
9. Project Management;
10. Quantity Surveying and Construction;
11. Residential – Real Estate Agency;
12. Residential – Residential Property Management;
13. Residential – Residential Survey and Valuation;
14. Valuation.

The choice of pathway must closely align to a candidate's role, experience and responsibilities. For example, on the Residential pathways candidates who are involved with HomeBuyer Reports and mortgage valuations are likely to be more aligned to the Residential Survey and Valuation pathway. On the other hand, those involved in block management and service charges may be more suited to the Residential Property Management pathway.

What Are the AssocRICS Competencies?

Each pathway requires candidates to demonstrate six technical competencies and eight mandatory competencies.

Candidates should download the relevant pathway guide from the RICS website, each of which has its own set of technical competencies. Candidates need to review these to ensure that they align with their role and experience, otherwise it will be impossible to demonstrate competence at level 2, i.e., 'doing'.

The relevant pathway guide sets out a clear description of each competency. This outlines the requirements that candidates need to satisfy, together with examples of skills, knowledge, experience and examples of relevant work-base tasks.

The technical competencies are the 'hard' skills required by candidates. Examples include Access and Rights Over Land for the Land and Resources pathway, Contract Practice for the Project Management pathway and Inspection for the Valuation pathway.

Some pathways will have optional technical competencies choices which candidates will need to select during the submission process. For example, Quantity Surveying and Construction candidates can choose between three optional competencies: BIM Management, Commercial Management of Construction, and Design Economics and Cost Planning. This allows candidates in a wide range of roles to tailor the AssocRICS assessment towards their specific requirements.

The eight mandatory competencies are 'soft skills' or business skills, common to all pathways. This means that they need to be demonstrated by all AssocRICS candidates, irrespective of role, sector or position. They aim to show that a candidate is able to work with others, can manage their own workload and act ethically and professionally.

The mandatory competencies are:

- Client Care;
- Communication and Negotiation;

- Conduct, Rules and Professional Practice – demonstrated by passing the RICS ethics module. No written statement is required for this competency.
- Conflict Avoidance, Management and Dispute Resolution Procedures;
- Data Management;
- Health and Safety;
- Sustainability;
- Teamworking.

What Does the AssocRICS Submission Involve?

The AssocRICS qualification is based on a written submission only. There is no face-to-face online interview as for the APC assessment. These means that it is vitally important for AssocRICS candidates to ensure that they present their submission to the highest written standards, including spelling, grammar and proofreading. It needs to be a 'client ready' document written in formal, professional language.

Candidates should ensure they take sufficient time to prepare for their written AssocRICS assessment. This is typically upwards of six months, providing time for the candidate to reflect upon their experience, prepare their draft submission and liaise frequently with their Counsellor. Rushing the process is not advised and generally leads to referral due to incomplete or unprofessional submissions.

AssocRICS candidates will need an allocated Counsellor and Proposer, who can be the same person. The Counsellor also needs to be a Chartered Surveyor (MRICS or FRICS) or an AssocRICS qualified surveyor with at least four years of post-qualification experience.

Typically, the Counsellor will be a candidate's supervisor, line manager or superior in their organisation. The role can also be outsourced providing that the allocated Counsellor is sufficiently familiar with the candidate, their role and their experience. The Counsellor will provide regular support to the candidate, as well as reviewing, providing feedback and signing off their submission via their own Counsellor ARC system.

The Proposer role involves confirming that the candidate is of the required standard, both professionally and ethically, to become an AssocRICS surveyor. This is easily confirmed by the candidate's Counsellor, which is the key reason why the Counsellor and Proposer roles are often provided by the same individual.

Candidates submit their submission to the RICS via ARC, including the following key submission elements:

- Summary of experience;
- Case study;
- CPD record;
- Ethics module and test.

The summary of experience is split into two core elements; six technical competencies and eight mandatory competencies. No supporting evidence, documentation or appendices can be submitted with the summary of experience, so candidates must ensure they address both their knowledge and examples within their competency statements. This should be based clearly upon the requirements of the relevant pathway guide and competency descriptions.

The technical competencies have a total word count of 2,000 words. This is absolute and a candidate will be referred if they exceed this by even one word. There is no 10% rule or similar, as candidates may have experienced within their academic studies. The word count can be allocated as the candidate wishes between each separate technical competency. It is generally sensible to allocate it relatively evenly between each to avoid any deficiencies arising in a candidate's knowledge and experience.

Each technical competency requires candidates to write a summary of their knowledge and work experience. This relates roughly to APC levels 1 (knowledge) and 2 (doing). Candidates should closely align what they write to the competency requirement, located in the relevant pathway guide, and include at least one, ideally two, examples for each competency. The examples should be related to a specific task or instruction and written in the first person and past tense, e.g., 'I worked on a rent review at X' or 'I worked under a JCT contract at X'.

The mandatory competencies have a total word count of 1,000, which is again absolute and can be split as a candidate wishes between each competency (seven in total). The Conduct Rules, Ethics and Professional Practice competency does not require any written submission and is solely assessed by the ethics module and test, so this is not included within the 1,000 word count. Candidates should again include ideally two specific examples of their experience against each mandatory competency.

Candidates also need to submit a case study report, which focusses on a single project or instruction that they have been involved with. This requires substantial involvement of the candidate in a variety of aspects, generally under the supervision of a Chartered Surveyor or other more senior professional.

The case study word count is 2,500, which is again absolute. Exceeding this constitutes grounds for referral and there is no leeway to exceed this. References to legislation, RICS guidance and case law can be included in the case study body and word count, but extensive quotations or reproductions should be avoided.

Appendices may be included to support the candidate's case study and these are not included in the word count. These may include graphics, calculations or plans, but only if they are relevant to the case study and referred to clearly in the text.

Candidates should choose a case study topic which focusses on two clear technical competencies, but that also demonstrates other relevant mandatory and technical competencies. The case study does not need to include aspects of every single competency, which is unlikely to be realistically achievable in any case. The case study should relate to the geographical area in which the candidate is assessed and should demonstrate a working knowledge of any specific legislation, caselaw or RICS guidance.

The candidate must have been involved in the case study project or instruction within the last two years, worked back from the submission date. If a project is out of date, then this is a point for referral. Some projects, e.g., a construction project or a site disposal, may extend over a period of more than two years. However, the candidate's main involvement must have taken place during the two year validity period. To demonstrate the timing of a candidate's involvement, a timeline could be included in the introductory section.

The RICS provides a case study template to download from ARC, including an introduction, approach, achievements and conclusion. This should be adopted by candidates, rather than choosing to use their own approach or structure.

The case study should include the following sections:

- Introduction/Context– Including a brief outline of the candidate's career history, role, the project, the candidate's supervision and the client's objectives and requirements. This effectively sets the scene and allows the assessors to understand the context and background of the project in terms of the candidate's experience and knowledge;

- The Approach – The candidate should discuss their role and responsibilities in the project or instruction, together with the knowledge and technical skills demonstrated. This should provide the assessors with a logical step-by-step journey through the process of the candidate working on the project or fulfilling the instruction;
- The Result – The candidate should discuss the outcome of the project or instruction, making clear references to their demonstrated competencies;
- Lessons Learnt – The candidate should reflect on their strengths, weaknesses and what they learnt and will improve on in similar future projects. The candidate should also include a conclusion to round off the case study in a similar manner to a professional report.

AssocRICS requires candidates to record at least forty-eight hours of CPD in the twelve months prior to submitting for assessment. This is calculated by the RICS on a rolling basis dating back from the candidate's submission date. Therefore, if a candidate submits in February 2021 then their CPD must be undertaken in the twelve months prior to this, i.e., from February 2020 to February 2021. The CPD requirements do not equate to the CPD recording requirements for qualified members, which are lower. This is often a common source of confusion for AssocRICS candidates.

All CPD logged by AssocRICS candidates should relate clearly to their technical and mandatory competencies, with relevance to the candidate's scope of work, role and responsibilities. Ideally, a candidate will undertake a variety of types of CPD to provide the widest possible learning benefits. Candidates should also plan their CPD in advance and ensure that they reflect on what they have learnt afterwards, demonstrating their analytical and evaluative skills.

At least 50% of the minimum forty-eight hours' CPD must be formal in nature. This requires a structured approach with clear learning objectives. Examples of formal CPD (RICS, 2020i) include structured seminars, professional courses (which do not have to be run by the RICS), structured self-managed learning and online webinars.

CPD is logged via ARC, including details of:

- Description of the Activity – Such as 'Webinar – JCT Contracts' or 'Self-managed learning on the Red Book';
- Activity Status – Future or completed;
- Start Date;
- Time – Hours and minutes allocated to the CPD activity;
- Activity Type – Formal or informal and method, e.g., webinar, conference, private study or mentoring;
- Learning Outcome – A candidate's reflection on what they learnt during the CPD activity. This should be written in the first person and past tense, e.g., 'I learnt about the different procurement methods and how they can be applied in my role' or 'I learnt about the new RICS Professional Statement and the core principles which I can apply to my involvement with commercial service charges'.

It is essential that candidates provide sufficient detail in their CPD record to allow the assessment panel to see how the CPD activity relates to their competencies, role and experience. If the CPD record is poorly written or incomplete, then it may constitute grounds for referral. The same applies if insufficient hours have been logged or if activities are informal rather than formal, meaning that the minimum requirements are not met.

Candidates also need to complete the online RICS ethics module and test via ARC, which provides sufficient evidence to satisfy the Conduct Rules, Ethics and Professional Practice mandatory competency (without any further written statement being required, as is the case for the other mandatory competencies). A candidate can, however, be referred on the Conduct Rules, Ethics and Professional Practice competency if their written submission, both in the summary of experience or case study, demonstrates a lack of understanding of competency requirements or there is evidence of questionable ethical or professional practices.

Candidates must have their employer's and client's express consent to disclose sensitive or confidential details in their submission. If this is not provided, then any specific details should be redacted. A candidate could state Project X instead of the address or use the term secured lending client rather than stating the lender's actual identity.

Anything that is litigious should also be redacted to ensure that any identifiable details are not disclosed, this could include the client's name, building address or sensitive financial information. Candidates can be referred for not dealing with confidentiality appropriately, as it is a key ethical consideration. The RICS also make a commitment to candidates that anything in an AssocRICS submission will not be disclosed further than the assessment panel.

How Do Candidates Use the Assessment Resource Centre?

Candidates need to use the Assessment Resource Centre (ARC) to start, monitor and conclude their AssocRICS assessment journey. ARC is also used by the candidate's Counsellor for the process of supporting and signing off the final written submission.

ARC is an online RICS system which is used to upload and submit the candidate's submission. Both the candidate and their Counsellor will have their own log-in and ARC system, but some of the information is shared between the parties, e.g., summary of experience and case study.

The candidate has their own home page on ARC which provides an overview of their AssocRICS progress. This includes the candidate's profile, candidate details, Counsellor details, progress overview and competencies. Changes to these can be made via the profile edit view, with the exception of the candidate's name, which needs to be notified to the RICS by email.

The candidate will need to contact the RICS to confirm any change in Counsellor details, as it is not possible to edit these directly via ARC. When nominated, the candidate's Counsellor will have access to the candidate's details via their own ARC system.

The candidate's profile needs to be filled out as soon as a candidate is enrolled and has access to ARC. This requires a photograph of the candidate, contact details, pathway, qualifications (academic and professional), employment history and Counsellor number.

The candidate's photograph should be professional and recent (within six months of applying for assessment), as it will be included on the front page of the final written submission. Rather than using a cropped photograph or one from a social occasion, candidates should use a simple professional headshot taken using a mobile phone, tablet device or digital camera against a white background or wall.

The candidate's final submission will include details of their qualifications and employment history, so it is important to ensure that these sections are filled out accurately and fully. Missing information will affect the assessors' abilities to review the candidate's final submission. The candidate will need to search for their employer on the RICS database and if they are not listed, the candidate will need to email the RICS to request their employer is added. The same may apply to academic or professional qualifications which are not already listed on ARC.

Any change that a candidate needs to make to the information held on ARC that requires input or action by the RICS, e.g., Counsellor or employer additions, will take time to be resolved. It is, therefore, recommended that any changes are made well in advance of the candidate's submission deadline. The Counsellor will also needed to have undertaken the RICS Counsellor training before they are able to sign off their first candidate via ARC.

The candidate will need to select their competencies on ARC using the competency selector view. This includes confirming the mandatory and core competencies and selecting the candidate's relevant optional competencies, if available. A candidate can edit their competency choices at any time before submitting via ARC, so these choices are not set in stone until the final submission is sent to the RICS. This may be relevant if a candidate's role or responsibilities change during the course of their AssocRICS journey or if they later feel that another competency choice would be more reflective of their day-to-day experience.

When preparing and uploading the written submission, the candidate will need to:

- Upload each separate competency wording to the individual competency boxes on ARC;
- Upload their case study in PDF format onto the case study view page. There is a Word Document template to download that candidates should use, which will then need to be converted to PDF format (using Adobe Acrobat, the Word export function or an online convertor) prior to uploading to ARC. Any appendices, images or tables must be included within one single PDF, as multiple documents cannot not be uploaded. The case study can be uploaded multiple times. If further changes are made only the most recent version will be included within the final submission;

- Upload their CPD to the CPD record view. This requires each individual CPD activity to be uploaded with a description of the activity, activity status (completed/planned), start date, hours and minutes allocated, activity type (formal/informal) and learning outcome. Of particular importance is ensuring that the learning outcome is sufficiently detailed and written in plain, clear and accurate English;
- Undertake the online ethics module and test within twelve months prior to assessment;
- Candidates should pay particular attention to the word counts for each section and ensure that they do not exceed these, as this will constitute a point for referral.

The Counsellor and Proposer sign-off process can be challenging and time consuming, so it is recommended that this is prepared for and undertaken in good time prior to the submission deadline. This process requires the following actions:

- Candidate to upload their competency wording and case study via their own ARC system. They then need to click the request review button in order for their Counsellor to review and either approve or provide feedback on each element of their submission;
- When the individual elements of the submission have been approved, the Counsellor will need to approve the candidate for final assessment using their ARC system;
- When the Counsellor has approved and proposed the candidate for final assessment and provided their Counsellor declaration (or a separate Counsellor and Proposer have approved or proposed the candidate), the candidate will note that their ARC screen shows green completed boxes next to each element of their submission;
- The candidate can then apply for final assessment when the submission window opens. This is usually confirmed by a banner at the bottom of the candidate's ARC window confirming the relevant submission window.

The Counsellor should be notified in the messages view when the candidate's submission elements are ready for review. However, these are not followed up by email notification from the RICS, so the Counsellor and candidate should liaise closely to complete the sign-off process within the required timescales.

When the candidate is able to apply for final assessment, they have a variety of further choices and options to select via ARC:

- International Experience – If relevant, the candidate should confirm this here;
- Special Considerations – including any extenuating circumstances or medical conditions that the assessors should be aware of, e.g., dyslexia;
- Download Submission – the candidate should download their PDF submission to ensure that they are happy with the final document and check carefully their proof reading, content and word count. ARC may make formatting adjustments in the spacing based on the candidate's word count, which may affect the presentation of the final submission. Unfortunately, there is no way to amend this and the assessors will be aware that this may be the case. The candidate will, therefore, not be penalised for any formatting or spacing issues relating to ARC in their written submission.

The candidate will then need to confirm they are ready to submit for assessment. When complete, this will be confirmed by a successful confirmation banner at the bottom of the page.

How Is AssocRICS Assessed?

The written submission is reviewed by a panel of two trained AssocRICS assessors. They will specifically refer to the pathway guides and competency descriptors when reviewing the candidate's submission. Therefore, it is essential that these are adhered to when a candidate is writing their summary of experience and case study.

The assessors will review the written submission holistically, ensuring that the requirements of the mandatory and technical competencies are met. The broad depth and breadth of knowledge at level 1 should be detailed in each competency summary of experience statement. However, candidates are unlikely to be able to demonstrate every single area of experience or task outlined in the competency guide. This is taken into consideration by the assessment panel. They will want to see that candidates can relate and apply their knowledge to their experience and can also clearly articulate this within practical, specific examples at level 2.

Specifically, when assessing the summary of experience, the assessors will be looking for the following:

- Strong technical knowledge which reflects the requirements of the pathway guide and competency descriptors;
- Evidence that the candidate has applied their technical knowledge through specific examples relating to their day-to-day work. This does not require candidates to provide independent client advice and can include assisting and shadowing a more senior surveyor. For example, preparing elements of a draft report, carrying out due diligence or assisting with an inspection. The standard required of AssocRICS is, therefore, clearly differentiated from the level required by APC candidates in terms of the provision of level 3 reasoned advice to clients;
- High standards of professional written work, including careful proof reading and use of formal language. A candidate's submission should be of the standard required for a client-ready report;
- Evidence that the summary of experience accurately reflects the candidate's role and responsibilities.

In the case study, the assessors will be looking for the following:

- A well-structured case study with a clear introduction, main body and conclusion;
- Clearly identified aims and objectives;
- Technical and mandatory 'soft' skills which reflect the candidate's competency choices and the evidence provided in the summary of experience. This needs to include two key technical competencies, although others can also be demonstrated;
- Sufficient reflection on what the candidate learnt from the case study and how they will apply the lessons learnt in the future.

The assessors are also looking for high standards of written communication, with strong visual communication skills demonstrated within the case study appendices, figures and tables. The submission must be professionally organised and presented, showing clear learning outcomes from the experience that candidates have gained during their AssocRICS journey. Candidates should also demonstrate their skills in analysis, reflective thought and problem solving throughout their summary of experience and case study.

If a candidate is successful, the RICS will confirm the successful result via ARC on the last working day of the month following the candidate's submission deadline. No additional feedback will be provided by the RICS in this eventuality. After qualifying as AssocRICS, candidates will have to continue to record at least 20 hours' CPD each year, of which at least 50% must be formal.

If a candidate is referred, the RICS will issue a referral report outlining the assessors' decision and reasons for the referral. Referred candidates have the right to appeal, although this cannot be on the basis that the candidate disagrees with the assessment panel's decision. Grounds for appeal, therefore, must only relate to a deficiency in the procedure or process of assessment which led to the referral.

As AssocRICS requires only a written submission, the RICS carry out both desktop audits and verification interviews for quality assurance purposes. These aim to ensure that candidates have undertaken the required experience and tasks and that their submission is solely their own work (and not that of someone else).

If a candidate is selected to undergo a desktop audit, the RICS will request additional supporting evidence or commentary from the candidate and their Counsellor in relation to the candidate's experience, role and responsibilities.

The online verification interview involves an auditor intensively interviewing the candidate to ensure that the written submission was purely their own work and not written by another third party. The auditor will not, however, re-assess the competence of the candidate.

All AssocRICS submissions will be assessed for plagiarism using the Turnitin system. This compares the similarity of the written submission to other candidates' submissions and also to other published written work. A verification interview may, therefore, be conducted at the request of the assessment panel if plagiarism is suspected.

In the event that either quality assurance process flags up potential issues, the RICS will investigate the matter further. In particular, plagiarism is a very serious offence and could lead to disciplinary proceedings by the RICS, being a clear contravention of the ethical standard to act with integrity.

If a candidate is referred, then they can use their own work when resubmitting and this will not constitute plagiarism. Candidates can also use appropriate wording from their AssocRICS submission towards a future APC submission and this will not constitute plagiarism. The requirements of the two assessments are very different so candidates should not be using any wording, structure or content verbatim.

Conclusion

In conclusion, this chapter has provided an overview of the AssocRICS submission and assessment process, in addition to setting out useful advice for seeking success. Later chapters discuss the referral and appeal processes in more detail, as well as discussing the role of third parties supporting candidates, e.g., the Counsellor.

6 Referrals

Introduction

In previous chapters, we looked at both the APC and AssocRICS assessments, together with advice and tips for seeking success.

However, success is not guaranteed and some candidates will find themselves in the position of being referred. In this chapter, we look at the referral process, how to deal with being referred and what to do next.

How Does the Referral Process Work for the APC Assessment?

If an APC candidate is referred, i.e., they did not pass their APC final assessment interview, the RICS will issue a written referral report by email within twenty-one days. This will provide guidance on what went well, what did not go so well and how the candidate can prepare for success in their next attempt.

A referred candidate can submit their written submission again at the next available submission window. However, candidates may decide to wait for a longer period if they feel that they need to gain additional experience or knowledge, to address deficiencies identified in the referral report.

Before resubmitting, candidates may need to do one or more of the following:

- Gain additional experience, as detailed above;
- Undertake targeted CPD to address concerns identified in the referral report, as well as continuing to undertake CPD to meet

DOI: 10.1201/9781003156673-6

the requirements of the RICS. This should include formal CPD to meet the 50% minimum requirement;
- Continue to record their ARC diary, if they are undertaking structured training;
- Consider whether the case study needs to be amended or replaced depending on the feedback provided by the assessors. Candidates should also check the date of their case study to ensure that it remains within the twenty-four month validity period set by the RICS;
- Update their summary of experience to include any new experience or examples, addressing any deficiencies identified in the assessors' feedback.

Candidates should review their referral report with their Counsellor (and Supervisor, if appointed) to discuss the assessment panel's feedback. This should form the basis of an action plan to address the concerns identified and ensure that the candidate is ready to resubmit at a future assessment window. The candidate's referral report is solely for their own use and will not be made available or shared with any future assessment panels.

When a referred candidate resits their final assessment interview, they will be considered as a first-time candidate by the RICS. This means that the candidate's assessment panel will not be aware that the candidate has been previously referred and the panel will not be provided with a copy of the candidate's referral report. This ensures that the assessment process remains fair and transparent, giving each candidate an equal opportunity to succeed at each assessment attempt.

How Does the Referral Process Work for the AssocRICS Assessment?

After a candidate submits their AssocRICS submission, a decision will be issued by the RICS within one month of the submission deadline. In the event of a referral, a comprehensive written feedback report will be provided to the candidate explaining the reasons for the referral. It will also contain advice on meeting the requirements in the candidate's resubmission attempt.

Before resubmitting, candidates may need to do one or more of the following:

- Gain additional experience to address concerns identified in the referral report;
- Undertake targeted CPD to meet concerns identified in the referral report, as well as continuing to undertake CPD to meet the requirements of the RICS. This should include formal CPD to meet the 50% minimum requirement;
- Consider whether the case study needs to be amended or replaced depending on the feedback provided by the assessors. Candidates should also check the date of their case study to ensure that it remains within the twenty-four month validity period set by the RICS;
- Update their summary of experience to include any new experience, or examples, addressing any deficiencies identified in the assessors' feedback.

Candidates should review their referral report with their Counsellor to discuss the feedback provided by the assessment panel. This will form the basis of an action plan to address the concerns identified and ensure that the candidate is ready to resubmit at a future assessment window. The candidate's referral report is solely for their own use and will not be made available or shared with any future assessment panels.

When a referred candidate resits their final assessment interview, they will be considered as a first-time candidate by the RICS. This means that the candidate's assessment panel will not be aware that the candidate has been previously referred and the panel will not be provided with a copy of the candidate's referral report. This ensures that the assessment process remains fair and transparent, giving each candidate an equal opportunity to succeed at each assessment attempt.

This means that candidates will need to resubmit their entire submission and the entire document will be subject to fresh scrutiny. Any competencies that were 'passed', i.e., successful, in the last submission will be reassessed by the new panel. This means that any updates to legislation, RICS guidance and hot topics need to be incorporated by the candidate in their resubmitted documents. There is also no guarantee that a competency or case study that was successful in the last referred attempt will be successful in the next attempt. This decision is down to the new assessment panel and subject to individual scrutiny and subjectivity. There is no longer the ability to 'bank' successful competencies, which was the position previously set by the RICS.

What Are Common Areas for Referral?

The pass rate for the APC typically varies from 60–70% year on year and between pathways.

The APC assessment is considered holistically; it is not an examination, and a candidate is not expected to be able to answer every single question correctly. Candidates are assessed against the standard required of an 'average' Chartered Surveyor. This is a professional who is a safe pair of hands, able to advise and liaise with clients effectively and who could, in theory, open their own RICS regulated Firm after qualifying.

This means that a candidate who broadly responds to most questions professionally and accurately, without any ethical concerns, should succeed in their final assessment interview.

The only area for an automatic referral is an incorrect answer on an ethics question. This would call into question the integrity of the candidate and cannot be considered holistically by the panel. Alongside an ethics referral, the assessors will raise any other minor or major concerns which arose during the interview or within the candidate's written submission.

At AssocRICS, the nature of the written submission means that the assessors are purely basing their decision on what the candidate has written. A referral may be based on a variety of concerns relating to the case study, summary of experience and relevance to the candidate's pathway and competency choices. Referral concerns may also relate to the standard of written English, proofreading, structure, style and professional presentation of the written submission.

Reasons for referral vary widely, including general concerns such as:

- Insufficient experience demonstrated in the examples (or lack of examples) stated in levels 2 and 3;
- Incorrect, inaccurate or inadequate knowledge stated in level 1;
- Lack of awareness of the requirements of the competency descriptions demonstrated across all levels;
- Case study which does not meet the requirements of the RICS, e.g., date validity, competency relevance or structure;
- Insufficient CPD hours, formal content, or relevance of activities undertaken to the candidate's chosen competencies and pathway;
- Lack of awareness of CPD activities when questioned on this element of the written submission;

- Poor presentation or lack of attention to meeting the ten minute timing requirement for the case study presentation;
- Poor responses to questioning, including lack of coherence, relevance or accuracy;
- Informal, unprofessional attitude, rather than considering the panel to be the candidate's 'client' and treating them accordingly;
- Poorly written submission, including lack of awareness of the competency guides, lack of compliance with the RICS requirements and poor proofreading and presentation;
- Lack of preparation leading to unfamiliarity with stated examples, experience and knowledge;
- Poor visual aid which detracts from the case study presentation;
- Poor awareness of ethical issues, RICS guidance and legislation, e.g., 5th Anti Money Laundering Directive, Bribery Act 2010 and changes to RICS regulation;
- Overall poor performance, which is often due to nerves, stress or a lack of confidence.

A referral will generally be based on a variety of issues, including an automatic referral decision if ethics was the principal cause. There are also many specific reasons for referral relating to the individual pathways and competency choices. Some of the most common are outlined below.

Building Surveying:

- Building Pathology is a competency where many candidates are referred. This typically relates to poor knowledge of dry and wet rot, rising damp and types of meter used to assess damp. Candidates need to ensure that they can explain the process of following the trail, diagnosing defects and advising on remedial action to clients;
- Candidates need to ensure that their case study choice is considered and suitable. Many referrals stem from a lack of reasoned advice and personal involvement in the case study project, with over-reliance on input or advice from contractors, superiors or other professionals. A failure to meet the requirements of level 3 in the relevant competencies could lead to referral;
- Lack of awareness of new or updated guidance, such as the Home Survey Standard or RICS Property Measurement (2nd Edition).

Commercial Real Estate:

- Candidates are often referred due to a lack of knowledge of VPS 2 of RICS Valuation – Global Standards (Red Book Global) and how this relates to inspections, limitations and desktop due diligence;
- Covid-19 has presented many challenges, particularly in relation to safe inspection procedures, health and safety, property management and business rates. Candidates need to be aware of relevant hot topics to ensure they are providing clients with the highest standards of level 3 reasoned advice;
- Landlord and Tenant is another common area for referral, particularly where candidates lack awareness of the fundamentals and principles relating to both lease renewals and rent reviews. Candidates also need to be aware of the Landlord and Tenant Act 1954, including Section 25, 26 and 27 notices and key timescales relating to these. Candidates should avoid referring to experience relating to property management or lettings in this competency, which should be referred to in the appropriate competencies;
- Candidates need to ensure that they refer to relevant experience in each competency, e.g., not confusing the requirements of Landlord and Tenant, Leasing/Letting and Property Management;
- Candidates need to be aware of the breadth and depth of the Property Management competency, including relevant RICS guidance. This includes the Professional Statement Service Charges in Commercial Property (1st Edition), Professional Statement Real Estate Management (3rd Edition) and Guidance Note Commercial Property Management in England and Wales (2nd Edition). Candidates should also be aware of common lease terms, including the differences between sub-letting and assignment;
- Measurement is another area for referral, where candidates are not aware of the requirements of RICS Property Measurement (2nd Edition). This specifically relates to the mandatory application of the International Property Measurement Standards (IPMS) to office and residential properties. Other IPMS standards, e.g., industrial and retail, have been published but not yet adopted by RICS in professional guidance. Candidates should be aware of and able to discuss the potential changes and differences

to the bases of measurement under the Code of Measuring Practice (6th Edition), which is incorporated within RICS Property Measurement (2nd Edition).

Quantity Surveying and Construction:

- Candidates often confuse the requirements of the Contract Administration and Contract Practice competencies. Minor referral points also relate to awareness of various contract types, e.g., JCT, FIDIC and NEC 1, 2 and 3, use of JCT Design and Build contracts and differences between cost reimbursement and lump sum contracts;
- Candidates need to be aware of current hot topics and the impact on their professional advice, e.g., advising on the impact of Covid-19 on contract clauses and project delays;
- Similarly to the Building Surveying pathway, candidates need to ensure that their case study choice is considered and suitable. Many referrals stem from a lack of reasoned advice and personal involvement in the case study project, having relied overly on input or advice from contractors, superiors or other professionals. A failure to meet the requirements of level 3 in the relevant competencies could lead to referral. This is particularly true in relation to the Commercial Management competency and when undertaking Cost Valuation Reconciliations (CVR);
- Referrals often relate to the Procurement and Tendering competency, where candidates are unable to explain the differences between procurement and tendering and the use of one or two stage tendering processes;
- Another common area of referral is on the Quantification and Costing of Construction Works competency. Knowledge-based deficiencies often include not being able to explain the differences between defined and undefined lump sums and the use of BCIS Standard Elemental Costs Plans.

Rural:

- Rural candidates are often referred on the Valuation competency, particularly in relation to their knowledge and application of RICS Valuation – Global Standards (Red Book Global). This includes understanding and being able to explain concepts such as yields, marriage or hope value, special purchasers and the use of the material uncertainty clause. Candidates should also be aware of the specific requirements of RICS Guidance Note Valuation of Rural Property (3rd Edition);

- In relation to the Landlord and Tenant competency, referral concerns often relate to knowledge of the Landlord and Tenant Act 1954, Agricultural Holdings Act 1986, and residential tenancy types;
- Candidates undertaking the Agriculture competency need to be aware of Statutory Management Requirement 13 relating to the welfare of farm animals and proposed changes to the Basic Payments Scheme (BPS) under the Agriculture Bill.

Valuation:

- Valuation can be a challenging competency for many candidates, with a wide variety of reasons for referral. These include the differences between VPS 4 Market Value and Fair Value, the differences between a development appraisal and a residual valuation and the appropriate bases of value to use for taxation valuations (e.g., Capital Gains Tax and Inheritance Tax). The latter stem from legislation rather than from VPS 4 of the Red Book Global, with full guidance being provided in the UK National Supplement;
- Candidates should be aware that the UK National Supplement should be read alongside the Red Book Global, rather than replacing it for UK-based valuation work;
- Candidates should be aware of the five methods of valuation and the three approaches to valuation as defined in VPS 5 of the Red Book Global. Although candidates are unlikely to have experience of all five valuation methods, they should know the theory and methodology behind each;
- Many candidates struggle with the fundamental principles of investment valuations. Specifically, this includes the use of different types of implicit and explicit yield, calculation of purchaser's costs and calculation methodologies such as term and reversion and hardcore and top slice;

- Other referral concerns relate to various competencies already discussed above, specifically in relation to the Commercial Real Estate pathway.

Planning and Development:

- Planning and Development candidates must undertake Valuation as a core competency, which is often a cause for referral due to insufficient experience or basic understanding of the level 1 fundamentals. Candidates should refer to the Valuation pathway concerns above for further common referral concerns;
- In relation to planning policy knowledge, candidates are often referred for not being aware of how listed buildings and conservation areas are treated in terms of policy and process. Candidates also need to be able to explain the application and differences between the Community Infrastructure Levy (CIL) and Section 106 agreements;
- Candidates should be aware of planning hot topics, such as the Planning White Paper and the changes to the planning use classes in 2020;
- Candidates are also often referred on the Legal/Regulatory Compliance competency, for example understanding and application of the Estate Agency Act 1979 and Construction (Design and Management) Regulations 2015.

Project Management:

- Generally, Project Management candidates need to ensure that they demonstrate sufficient understanding of the contract process, the process of novation and the wider role of the Project Manager across the project lifecycle;
- Candidates are often referred for a lack of understanding of the Construction Technology and Environmental Services competency. This includes knowledge and application of the RIBA Plan of Work 2020 to their practical experience;
- Similarly to Quantity Surveying and Construction candidates, referrals often relate to the Procurement and Tendering competency. Candidates are often unable to explain the differences between procurement and tendering, the use of one or two stage tendering processes and how tendering works under various contract types, e.g., JCT.

Residential:

- Candidates are often referred on the Valuation pathway, as their experience and knowledge is often limited to the comparable method only. Candidates should be able to discuss a wide variety of valuation methods, including practical application of more than one method. This could be via application of the investment method to buy to let properties or the residual method to land with development potential.
- Candidates often give poor explanations of how to value protected tenancies or Houses in Multiple Occupation (HMOs), in addition to not being able to explain differences between leasehold and freehold tenure in valuation terms;
- Candidates often struggle with the principles of valuation relating to leasehold enfranchisement, including how to assess relativity and apply precedents set by the Sloane Stanley Estate v Mundy (2016) case;
- Legal/Regulatory Compliance is another common area for referral, with a lack of awareness of legislation such as the Homes (Fitness for Human Habitation) Act 2018, Tenant Fees Act 2019, and future changes relating to the Minimum Energy Efficiency Standard (MEES).

How Can Candidates Deal with the Emotional Aspects of Being Referred?

Being referred is a stressful and upsetting experience, particularly given the importance of the final assessment interview as the last hurdle in a long journey to becoming a Chartered Surveyor. This will inevitably lead to a very emotional reaction from referred candidates. Being able to deal with this calmly and logically will put candidates in the best possible position to tackle the referral concerns, resubmit their documents and succeed at a future assessment interview attempt.

After receiving the referral decision, candidates should take some time to reflect on, and digest the outcome. This might mean taking a week or two out from the APC experience and focussing on spending quality time with family and friends. These are relationships which might have been neglected during the pressurised time leading up to the final assessment interview and now need some attention.

Candidates may also wish to speak in confidence with the RICS LionHeart charity about how the referral has affected them. Various ways to contact LionHeart are outlined at https://www.lionheart.org.uk.

LionHeart can provide counselling and support to candidates and this may be a useful support tool during challenging times. Candidates should never feel alone in their referral, as it will be experienced by roughly 30% of candidates sitting their final assessment interview. Reaching out to others about the referral experience can help – a problem shared really is a problem halved.

Candidates may also be served well by practising a mindfulness technique known as RAIN, introduced by psychologist Tara Brach. This involves recognising what is happening, allowing the experience to be there, investigating with interest and care and then nurturing with self-compassion. This can be a very useful tool to overcome negative emotions and feelings and to work towards a place of positivity, enabling the candidate to move forward.

Candidates will later receive their referral report from the RICS, which may trigger further distressing feelings and reactions. It is sensible for candidates to sit down with their Counsellor to discuss the content of the report and to agree an action plan to seek success at their next assessment attempt.

A minority of candidates may decide to appeal and this needs to be pursued within the timescales set out by the RICS. Candidates considering this should refer to a later chapter for a full explanation of the process. An appeal cannot be pursued because a candidate disagrees with the final decision of their assessment panel. It can only be pursued on the basis that the assessment process was not run fairly and transparently, for example if there were significant technical difficulties or the timing of the interview was not in line with the RICS requirements.

Candidates should try to rationalise the outcome and their referral feedback. Practical ways to do this include:

- Embracing emotions and going through the process of disappointment, frustration and sadness, for example. This will allow candidates to move to a more positive place when the time is right;
- Focussing on the referral as providing feedback for the future and a learning opportunity, rather than considering it to

constitute a failure. Candidates can also focus on the positives highlighted in the referral report and see the feedback as providing them with ways to improve as a professional;
- Putting an action plan in place to focus the candidate's mind on what they can do next to work towards becoming a Chartered Surveyor;
- Understanding that the interview process is subjective and the product of only one hour in front of a panel. This means that a 'bad day in the office' or suffering particularly from stress or nerves may have affected their performance;
- A candidate may also not agree with the outcome but understanding that the outcome was the combined decision of a panel or two or three Chartered Surveyors means that it should be a fair and balanced decision.

Candidates who substantially suffered from nerves or stress during the interview process should seek support to overcome these. This could be through techniques such as visualisation or mental rehearsal, as well as mock interviews or question and answer sessions. Candidates may also need to accrue further experience to boost their confidence, particularly in relation to how they are able to apply their level 1 knowledge to their level 3 reasoned advice to clients.

What Should Referred Candidates Do Next?

Any referred candidate should try to remain positive and implement a clear action plan to reach their desired goal of becoming a Chartered Surveyor. They should put in place positive steps to satisfy the referral concerns identified in their referral report and proceed to resubmit for final assessment when they feel ready. In some cases, this may not be at the next available opportunity, as candidates may feel that they need additional time to prepare or to gain experience in order to address deficiencies identified in their referral feedback.

Candidates who are referred should be aware of the six year rule, which requires candidates to pass their AssocRICS or APC assessment within six years of initial enrolment. A referred APC candidate who is nearing this deadline may wish to pursue the AssocRICS assessment as this will mitigate the requirement to meet the six year rule, allowing them sufficient time in future to develop their skills and knowledge to the standard required of a Chartered Surveyor.

Conclusion

In conclusion, being referred is an inevitably upsetting and disappointing process to go through. However, working through the referral feedback will help candidates to become even better Chartered Surveyors when they are successful at a final assessment interview. In the next chapter, we look at the appeals process which may be applicable to a minority of candidates where their interview process was not conducted fairly or appropriately.

7 Appeals

Introduction

In the previous chapter, we looked at how the referral process works for the AssocRICS and APC assessments. In this chapter, we look at the appeal procedure for candidates, potential grounds for appeal and how appeals are considered by appeal panels.

Who Has the Right to Appeal?

Any candidate who is referred in their final AssocRICS or APC assessment has the right to appeal. However, this does not mean that all referred candidates should appeal as there are limited grounds upon which an appeal can be mounted.

Candidates cannot appeal simply because they disagree with their assessment panel's decision to refer them; for example, because they were deemed not to be sufficiently competent to meet the requirements of the APC. Appeals are only considered on the grounds that there was a problem with the process, either administrative or procedural, leading to the referral. This will need to be on the balance of probability that the assessment was not conducted fairly or transparently.

When Should a Candidate Appeal?

Candidates considering submitting an appeal should first review their referral report, documentation and interview experience (if relating to an APC appeal) with their Counsellor, Supervisor and/or employer. This should include consideration of the following:

DOI: 10.1201/9781003156673-7

- Whether the candidate's written submission met the RICS requirements in terms of format, structure and requirements;
- Whether the candidate's presentation was within the ten minute time allowance and included relevant information (APC only);
- Whether the candidate addressed the panel's questioning adequately and if they did so in a clear, concise and efficient manner;
- Whether the candidate's questioning was relevant to their declared levels and competency choices;
- Whether the candidate's interview followed the RICS process in terms of structure, format and timings.

If a candidate feels that there were procedural or administrative issues that contributed to their referral, then this may provide sufficient grounds on which to mount an appeal.

In other instances, a candidate may decide that this was not the case and that improvements could be made to their submission or interview approach instead. In this case, candidates can reapply for assessment as a first-time candidate when they feel ready to do so at a future date.

There are no set grounds for appeal, however, the following are examples of where an appeal may be upheld:

- There was a conflict of interest between the chair, or an assessor and the candidate;
- The assessment panel members were provided with the wrong, or an incorrect written submission by the RICS;
- The panel's questioning focussed on areas outside of the candidate's declared levels and competency choices (APC only);
- The interview timing was too short, or too long, or there were significant technical difficulties that were not dealt with appropriately (APC only).

Who Considers Appeals?

The RICS instructs appeal panels to consider appeals relating to both AssocRICS and APC assessments. These are generally comprised of three Chartered Surveyors, who are also trained assessors. They would not have been previously involved with the candidate's final assessment and would have had no prior relationship with the candidate, their experience, or their employer.

The appeal panel will receive the candidate's appeal form, referral report, assessment panel's mark sheets, assessment panel's response and any auditor's report (if available).

How Does the Appeal Process Work?

If a candidate decides to appeal, they must submit their appeal within twenty-one days of receiving their referral report from the RICS. This requires completion of the AssocRICS or APC candidate appeal form, which can be downloaded from the RICS website, and payment of a £100.00 fee to the RICS.

A decision by the appeal panel should be issued within seven weeks from the date that a candidate's appeal form is received by the RICS.

What Does the Appeal Form Include?

The appeal form requires candidates to submit the following:

- Basic details, including name, employer and assessment details;
- APC – A clear, structured summary of the reasons for appealing. The word count is 1,000 words and this must not be exceeded;
- AssocRICS – A twenty word summary of the candidate's grounds for appeal followed by a summary of 1,000 words detailing the reasons for the appeal. Again, the word count must not be exceeded.

Candidates cannot include any appendices or supporting documentation with their appeal, such as any statements from their Counsellor, Supervisor or employer, for example.

What Happens if a Candidate's Appeal Is Unsuccessful?

If a candidate's appeal is unsuccessful, they will be notified by email from the RICS of the decision together with the supporting reasons. There is no further right to challenge the appeal panel's decision. The candidate can then reapply for final assessment when they are ready, as would be the normal case in the event of a referral.

What Happens if a Candidate's Appeal Is Successful?

If a candidate's appeal is successful, they will be notified of this by email from the RICS and will have the opportunity to undertake their final assessment again. The appeal panel cannot overturn the decision of the original assessment and pass the candidate based on their previous performance. However, the candidate can then discard their referral report, although it is advisable to take note of any areas for improvement highlighted.

The candidate will be invited for reassessment by the RICS within three months of the date of the appeal decision. Candidates will be allocated a completely new assessment panel and will be assessed again, based on their existing written submission.

Successful candidates will have their appeal fee held in credit by the RICS towards future fees.

Conclusion

This chapter has outlined the process involved with appealing the assessment panel's decision for both the AssocRICS and APC final interview assessments.

In the next chapter, we will look at the supporting roles relating to both the AssocRICS and APC assessments, including the role of the Supervisor and Counsellor.

8 Supporting Candidates

Introduction

Candidates looking to become a Chartered Surveyor do not need to undertake their professional journey alone. There are many people who will be able to support and provide assistance to candidates, both professional and personal, at the right times during the process.

In this chapter, we look at the various support roles relating to both the APC and AssocRICS qualifications. This includes advice on how candidates can access support and benefit best from what is being offered.

Who Supports Candidates Wishing to Become a Chartered Surveyor?

The RICS require that all candidates have a Counsellor during their AssocRICS or APC journey. This chapter will take an in-depth look at who can fulfil this role and what type and level of support should be provided.

Candidates can also access support from a Supervisor, RICS Matrics and other organisations who provide outsourced support to candidates. Candidates' employers should also ideally be involved in providing support.

Candidates will be supported during their final assessment interview for the APC by their assessment panel, generally comprising of a chairperson and one or two assessors.

Primarily, however, candidates need to be responsible for their own journey to become a Chartered Surveyor. They need to drive

DOI: 10.1201/9781003156673-8

their own process forward and take responsibility for seeking the right support at the right time. This requires a commitment to follow the process from the beginning to the end, bearing in mind the requirement to become MRICS within six years of enrolment (the six year rule).

Both the APC and AssocRICS processes can be stressful, challenging and demanding, and should only be undertaken with the right attitude and support in place.

What Is the Role of the Counsellor?

Every single APC and AssocRICS candidate must have an appointed Counsellor to provide them with support and guidance. The Counsellor is also responsible for providing the final sign off for the candidate's written submission, verifying that they have achieved the required levels of competence and have sufficient relevant experience and knowledge.

A candidate's Counsellor does not necessarily have to work with the candidate. They also do not have to have personal experience of all the candidate's chosen competencies. They may be the candidate's line manager or another senior colleague within their organisation. Alternatively, they may be external to the candidate's organisation, providing outsourced support where this is not provided in-house or in the instance of a self-employed candidate, for example.

For the APC, the Counsellor must be a Chartered Surveyor qualified either to MRICS or FRICS membership level. For AssocRICS, the Counsellor must be a Chartered Surveyor qualified either to MRICS or FRICS membership level or to AssocRICS level with at least four years' experience.

Counsellors need to complete mandatory face-to-face or online training provided by the RICS prior to supporting candidates. They should also consider completing the online ethics module and test, which is also a requirement for candidates to complete as part of their final assessment submission. This will ensure that the Counsellor has up-to-date knowledge of current ethical standards and issues and can provide the candidate with a high standard of support and guidance.

The level of support that a Counsellor provides should be tailored to the requirements of each individual candidate. This may be far more involved and technical for structured training candidates with less practical experience, compared to a Senior Professional candidate who may require more peer support and guidance primarily focussed on meeting the RICS requirements.

All Counsellors should be familiar with the RICS guidance relating to the APC or AssocRICS, including the Candidate, Counsellor and Pathway Guides. This includes familiarity with the mandatory and technical competencies, together with the structure and format of the APC and AssocRICS submission and assessment processes.

Counsellors and their candidates should work collaboratively together over the period of the candidate's APC or AssocRICS journey. This is primarily through quarterly, or more frequent meetings, as required, to discuss the candidate's progress and action plan for the coming months. These meetings can be in person or by video or phone call. Liaison will also be via the Assessment Resource Centre (ARC), to which both the candidate and Counsellor will have their own separate access.

Although the Counsellor is there to support and guide the candidate, it is the candidate's responsibility to be accountable for their own APC process. This includes the candidate driving forward the process and setting up meetings or drafting their documents in good time, for example.

The APC Counsellor's role will include some or all of the following responsibilities and tasks:

- Working with the candidate's Supervisor, if appointed, to provide consistent guidance and support;
- Working with the candidate to monitor, review and plan their work experience and CPD activities. This will include identifying any deficiencies or gaps in a candidate's experience or knowledge and planning ways to overcome these;
- Understanding the RICS guidance provided to Counsellors and candidates to provide the highest standard of support and advice;
- Helping the candidate to select the most appropriate route, pathway and competency choices, based upon their knowledge, experience and role;
- Helping the candidate to enrol on the APC, including filling out the relevant forms and registering on ARC;
- Registering their Counsellor account on ARC to liaise with their candidate and provide the final sign-off;
- Providing guidance, support and encouragement to their candidate throughout their journey to becoming a Chartered Surveyor;

- Meeting quarterly (or more frequently) to review and advise on progress and future objectives;
- Helping the candidate to select a suitable case study topic based on their competency choices and current projects or instructions they are involved in;
- Working with the candidate to collate their final assessment submission, ensuring that all relevant documentation and supporting evidence is included. This applies to both the preliminary review and final submission process. At preliminary review, there is even more onus on the candidate and Counsellor to ensure that the submission meets all of the RICS requirements;
- Reviewing the final submission with the candidate to ensure that it meets the requirements of the RICS and adequately meets the required competency levels. Careful proof reading should be undertaken and encouraged by the Counsellor to ensure that a professional 'client ready' document is submitted by the candidate;
- Signing off the candidate's submission only when the Counsellor feels that the candidate is ready to be put forward for final assessment. If the Counsellor does not consider that the candidate is ready, e.g., they have not gained sufficient experience in a particular competency, then they should discuss their reservations with the candidate. They should also provide guidance on how to address this in a constructive manner. Under no circumstances should a Counsellor sign off a candidate if they do not feel they are ready, even if they are under pressure to do this by the candidate or their employer;
- Providing advice and guidance on how to prepare most effectively for the final assessment interview, which may include a mock interview. To ensure candidates get the most out of this process, it is advisable for Counsellors to complete the RICS assessor training, or, at least, to review the assessor training guidance. This will mean that candidates receive a mock interview that is closely aligned to what they will experience at the real thing. Counsellors should also encourage candidates to undertake question and answer sessions and mock interviews with other trained assessors to give them as much experience as possible before they sit their final assessment interview;
- Providing support to the candidate if they are referred, including reviewing the candidate's referral report and putting together an action plan to address the feedback and referral concerns raised.

For structured training candidates, the Counsellor will also need to:

- Plan the candidate's period of structured training and monitor their progress as they complete, ideally surpassing the minimum number of days of experience required;
- Review the candidate's diary and logbook at least at the quarterly review meetings to discuss current progress against the required competency levels and overall standards required to become a Chartered Surveyor. Every half year, the candidate and Counsellor should work together to review the summary of experience and update this in line with new experience and knowledge gained. These meetings also give both parties the opportunity to identify problems or challenges and explore ways to overcome these;
- Ensure they have a good understanding of the candidate's progress in relation to the required competency levels. At level 2, this requires the candidate to competently undertake a task without Supervisor. At level 3, this requires the candidate to provide reasoned advice to clients, including analysis of various options and recommending the most viable.

The RICS provides a suggested timeline for Structured Training candidates, including the frequency and content of Counsellor meetings, in Figure 8.1.

For preliminary review candidates, the Counsellor will also need to:

- Help the candidate to identify how their past and current experience fits with their chosen competencies. If experience was gained some time ago, the Counsellor may need to work with the candidate to ensure they are familiar with the process and any updated or more current guidance or legislation relating to that particular area of practice;
- Ensure that the candidate's preliminary review submission is of the highest standard possible, addressing all RICS process requirements, as well as demonstrating the candidate's ethical, technical and professional capabilities in a well-written submission;
- Review the preliminary review feedback report, ensuring that the candidate is able to prepare in advance for their final submission and make any changes or amendments recommended by their assessment panel.

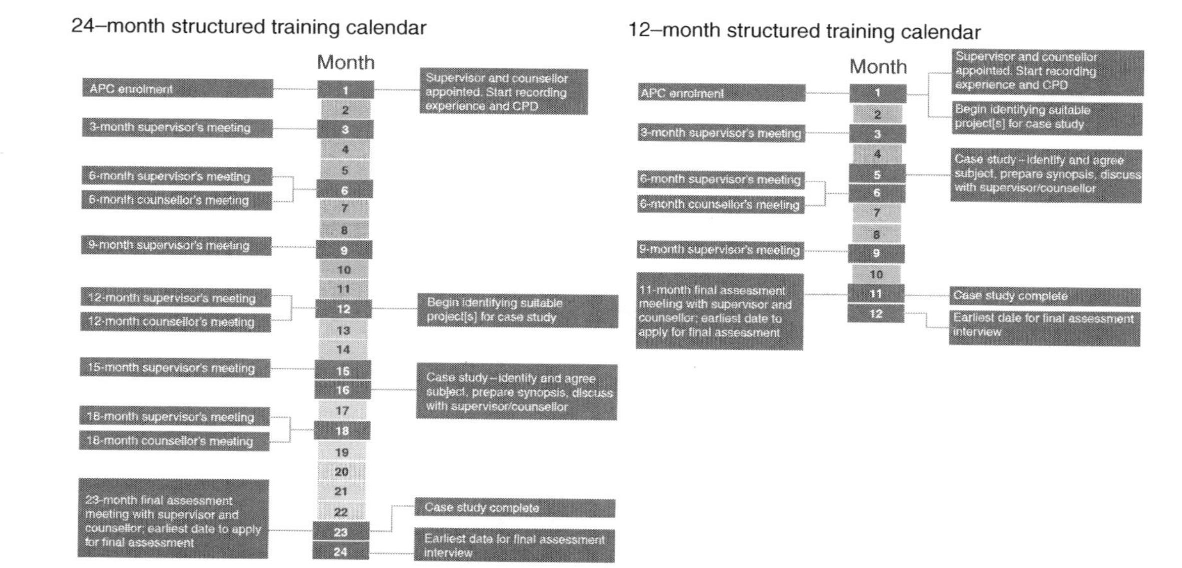

Figure 8.1 Timeline for Structured Training Candidates (RICS, 2020j)

Counsellors who do not work with their candidates on a day-to-day basis have an even more challenging role in ensuring that they provide a high standard of support and guidance. This will require excellent communication skills and constant collaboration to explore the following during regular meetings:

- How the candidate is performing in their day-to-day role, including soft skills such as working in a team, problem solving and time management;
- What the candidate is producing in terms of work-based outputs, including reviewing reports or instructions together and discussing how these relate to the chosen technical competencies;
- Why the candidate provided certain advice to a client, analysed specific options or justified their approach to a problem. Delving into the candidate's thought processes and logic will help to build a better submission, including choosing the case study topic and developing the key issues to be included within this;
- How the candidate's advice and role relates to their wider working environment;
- How the candidate is developing a professional and ethical attitude to becoming a Chartered Surveyor.

The role of the Counsellor working with Senior Professional, Specialist and Academic candidates will be slightly different. Primarily, the two parties may work together on a peer mentoring basis, with a focus on:

- Helping the candidate to relate their professional experience to the chosen competencies and requirements of the APC;
- Working with the candidate to understand the fundamentals and basic principles of their professional work. This can require going back to basics, which the candidate may have learnt many years ago during formal education. This often becomes second nature and can be difficult for the candidate to articulate in simple terms or in terms of first principles;
- Ensuring the candidate is aware and familiar with the highest standards of ethics and professionalism;
- Helping the candidate to structure their submission in line with the requirements of the RICS. For example, structuring the case

studies appropriately or including sufficient detail of the CPD learning outcomes.

Most importantly, the Counsellor has the responsibility to provide the candidate with the final sign-off via ARC. This declaration confirms that the candidate could undertake their current role and responsibilities in another organisation or firm, but that they could also set up their own regulated firm after qualifying. This confirms that the Counsellor is confident that the candidate is of the required competence to become a Chartered Surveyor and can act ethically and professionally in their day-to-day duties.

The requirements and role of the AssocRICS Counsellor and Proposer are slightly different. To be a Counsellor to AssocRICS candidates, the Counsellor must be an AssocRICS member with at least four years' experience, MRICS or FRICS qualified. This could be the candidate's line manager or Supervisor, but it could also be an outsourced party with sufficient familiarity with the candidate and their experience.

The AssocRICS Counsellor role includes:

- Being familiar with the candidate, Counsellor and relevant pathway guidance and being able to advise the candidate on how to meet these. This will include applying the candidate's experience to the competencies, selecting appropriate CPD activities and choosing a suitable case study topic. It may also include identifying gaps in the candidate's knowledge or experience that need to be filled through additional training, experience or CPD activities;
- Providing guidance and support to the candidate throughout the process. The RICS refer to this role as being a 'critical friend';
- Regularly meeting with the candidate, roughly every four to six weeks. These meetings should help to motivate and encourage the candidate, as well as providing critical momentum to continue moving towards the AssocRICS goal. A written action plan should be agreed and monitored, including six weekly competency progress reviews, three monthly overall progress reviews and six monthly strategic reviews;
- Providing support to draft, review and finalise the written submission.

The Counsellor is also likely to act as the candidate's Proposer. This requires the Proposer to declare that the candidate is of the standard required to become an AssocRICS member.

What Is the Role of the APC Supervisor?

APC candidates may also have an appointed Supervisor, although this is not a mandatory requirement. This is not a role relevant to the AssocRICS qualification.

A candidate's Counsellor may also act as their Supervisor. The Supervisor role encompasses day-to-day supervision and management of a candidate's role, including training and support particularly for structured training candidates. If a candidate does have a Supervisor, then they should liaise with the Counsellor to ensure that consistent guidance is provided to the candidate at all times. If inconsistent guidance is given, it can be very confusing to candidates, particularly in relation to competency choices and structure of their written documents.

How Do Assessment Panels Support APC Candidates?

A candidate's assessment panel fulfils the final supporting role during the final APC assessment interview. The panel members are there to ensure that an APC candidate is sufficiently competent, ethical and professional to meet the requirements of being a Chartered Surveyor.

However, the final assessment is an assessment of competence, not an exam. The assessment panel is there to ensure that the candidate is of the required standard, something which has already been verified and signed off by the candidate's Counsellor. The role of the panel is not to be adversarial and negative. The assessment panel is there to support the candidate through an interview via a process of questioning and assessment in relation to the candidate's declared competencies and written submission. The panel will act with integrity and professionalism and treat all candidates with due respect and courtesy.

Candidates should, therefore, view their assessment panel as a conduit to success. This requires candidates to work with their panel to answer questions diligently and accurately and to remain calm and positive towards the process.

How Can an Employer Support Their Candidates?

An employer is a key member of a candidate's support team. The support and training from employers can vary widely, from organised corporate training programmes in the largest surveying firms through to little or no support in niche organisations or if a candidate is working client-side or is self-employed. Employees will particularly be involved in the training and support of apprentice candidates, where the end point assessment is either AssocRICS or the APC.

That said, all candidates should be the driving force in their own AssocRICS or APC process. This means that they should take responsibility, being one of the five professional and ethical standards, to engage and work with their employer to support them to becoming a Chartered Surveyor. This may simply include the candidate engaging with an organised training programme. However, at the other end of the spectrum, it may require a candidate to actively organise and plan their own support programme through an external training provider.

An excellent tool for all candidates to plan their APC training and support is using the RICS Candidate Training Plan. This is particularly relevant for candidates undertaking structured training. However, it could easily be adapted to provide a self-assessment form for candidates on all other routes. The aim of the plan is to help the candidate and their employer work together effectively by understanding the requirements of the chosen competencies and the level and type of support required by the candidate. The plan includes the commitments of both parties towards one another, including issues such as payment of RICS fees, annual leave, CPD and referred candidate policy.

Candidates may find that their employer is unable to provide them with sufficient experience relating to all chosen competencies. This may require reconsideration of the chosen optional competencies. Alternatively, a candidate in a niche or narrow role may need to secure a temporary secondment or additional work experience to fulfil any deficiencies in their experience or knowledge.

For example, some Commercial Real Estate candidates in a niche role or small firm may struggle to gain adequate valuation experience. This may require them to be seconded into another firm to gain the required experience and supervision to meet the requirements of this core technical competency (to level 2). Candidates in this position should speak to their employer to identify any gaps in their experience and put an action plan in place to address them. Most employers will be open to working with the candidate to do this and may even have experience of similar situations being resolved for past candidates.

Candidates should always speak out, if possible, to overcome challenges such as this. Being open and honest with their employer will help them to succeed in the AssocRICS or APC process.

What Other Organisations Support Candidates?

Candidates can access support from a variety of other organisations, including RICS mentors, RICS Matrics, their university, RICS LionHeart and external training providers, such as Property Elite.

This may be particularly relevant if a candidate is working in a firm or organisation where there is no formal RICS APC or AssocRICS support. Typically, candidates will find themselves in this position if they are working client-side, are self-employed, or work in a very niche sector or market.

Candidates can certainly become a Chartered Surveyor in any of these environments, however, to improve their chances of success they may wish to consider seeking external support. This is particularly important given that candidates need to ensure they comply with the RICS requirements. They also need to ensure that they can demonstrate their competence to the required levels and to prepare adequately for the final assessment submission and interview (APC only).

RICS mentors are volunteer Chartered Surveyors who provide independent support to candidates and their Counsellors. The role is distinct from being a Counsellor and provides supplementary support throughout the process. A mentor should also not be expected to fully review or check a candidate's submission prior to assessment.

A candidate does not have to have a mentor as part of their journey to becoming a Chartered Surveyor. However, seeking the advice of a mentor aligned to a candidate's specific role or area of practice may provide extremely useful guidance and support. Advice from a mentor could cover some of the following:

- Aligning a candidate's training and experience to the required competency levels;
- Structuring and writing a submission, including case study topic choice, which meets the requirements of the RICS;

- Preparing adequately for the final interview, including presentation and questioning practice.

RICS Matrics provides a valuable support role to both students and newly qualified Chartered Surveyors. Access to Matrics is inclusive and is not based on age or a candidate's stage in their career. There are approximately forty UK Matrics bodies which run a variety of events, both online and in person, for both CPD and networking purposes. This may include advice on the AssocRICS and APC processes, as well as providing opportunities to meet and collaborate with other Chartered Surveyors and built and natural environment professionals.

Universities are also a source of additional support for students, apprentices and placement students undertaking their first twelve months of structured training. They may provide support on technical issues relating to specific competencies, professional ethics, soft skills such as confidence in presenting to clients, CPD opportunities and careers guidance. For apprentices and placement students, in particular, maintaining a good dialogue and collaboration between the university and employer is essential to providing a joined up and constructive journey to becoming a Chartered Surveyor.

RICS LionHeart is an RICS charity providing support to all RICS professionals, including financial support, counselling, work-related support and legal advice. LionHeart runs a variety of webinars, as well as providing confidential one-to-one support.

Finally, there are a variety of external training organisations, such as Property Elite (co-founded by the author), which provide training and support to AssocRICS and APC candidates. For some candidates, particularly those without any formal in-house support, this may include the Counsellor role. Candidates wishing to benefit from additional support from the beginning to the end of their qualification journey should make contact directly with these organisations.

Conclusion

In this chapter, we have looked at the various support roles relating to the AssocRICS and APC qualifications, including the Counsellor. This is the most important support role for both qualifications and is essential to provide support and guidance to candidates wishing to become Chartered Surveyors.

In the next chapter, we look at the next career steps that a Chartered Surveyor may wish to take, starting first with the FRICS qualification.

9 FRICS

Introduction

After becoming MRICS, often Chartered Surveyors look to the next step in their career progression. For many, this will involve looking to become a Fellow of the RICS (FRICS).

In this chapter, we will look at the process, requirements and top tips for candidates to be successful in their first submission attempt.

Who Can Become a Fellow of the RICS?

Any Chartered Surveyor who is MRICS qualified can apply to become a Fellow of the RICS. However, it is not an automatic progression after a specific minimum time period.

Whilst passing the APC will have demonstrated a candidate's range of relevant technical and mandatory competencies, some to the standard of providing reasoned advice to clients (level 3), FRICS requires candidates to do much more than just show a minimum level of competence.

FRICS is an 'honoured class of membership awarded on the basis of individual achievement within the profession' (RICS, 2020m). This requires candidates to demonstrate enhanced skills, career progression, recognition and profile within the industry. This sits alongside continued professional progression and promotion of the aims of the RICS for the public advantage.

Becoming a Fellow of the RICS, therefore, allows a Chartered Surveyor to stand out from the crowd. It is an outward marker of professionalism and excellence to colleagues, peers, the industry, clients and the wider public.

DOI: 10.1201/9781003156673-9

Fellows (FRICS members) are required to adhere to three core principles set out by the RICS:

- To further the RICS and surveying profession;
- To act for the benefit of third parties to reflect the RICS' public interest mandate;
- To promote the RICS' objectives and the wider profession.

If a Chartered Surveyor has any outstanding conduct issues on their RICS record, they will be unable to apply to become a Fellow.

How Does the Process to Become a Fellow Work?

FRICS applicants need to submit a written application form to the RICS, along with the appropriate assessment fee. Upon successful qualification, Fellows must pay an increased annual subscription fee to the RICS.

The FRICS application form requires the following to be completed:

- Personal details;
- Personal statement on current role, practices and ambitions (500 words);
- Employment history, including employer, position/job title, key dates and scope of responsibilities, for all relevant roles starting with the most recent;
- Academic qualifications, including subjects, qualifications and key dates;
- Professional body memberships, including subjects, qualifications and key dates;
- Written statements (500 words for each) and descriptions of third party evidence for each of four chosen professional characteristics;
- Referee, who must be MRICS and FRICS, who may be contacted by the RICS for confirmation that any information provided within the FRICS application is accurate;
- Name to appear on the Fellowship diploma;
- Declaration of truth.

At the time of applying, applicants must ensure that their online CPD record is up to date.

Once the application form is submitted to the RICS, the assessment process generally takes up to eight weeks. This may be delayed if a candidate's application is incomplete or the RICS need to investigate any characteristics or supporting evidence in further detail. This means that it is vitally important to attach comprehensive, relevant third-party supporting evidence to document each of four selected professional characteristics.

The decision may also take longer for MRICS Chartered Surveyors who have been qualified for fewer than five years. This is because the candidate's application will be reviewed by the relevant RICS World Regional Board or a group with delegated authority.

If an application is unsuccessful, the RICS will issue a feedback report explaining the reasons for this. In some instances, the RICS may request additional supporting evidence to assist with consideration of a candidate's applications.

In the event of an unsuccessful application, there is the right to appeal which can be pursued in the event of a procedural error. However, candidates cannot simply appeal because they disagree with the decision of the RICS assessment panel.

What Are the Professional Characteristics that Fellows Need to Demonstrate?

The RICS set out twelve professional characteristics, of which candidates need to demonstrate and select four within their Fellowship application. These are divided into four categories: Champion, Expert, Influencer and Role Model. Each of these four categories is split into three separate characteristics.

The Champion category relates to how a candidate has gained recognition from an appropriate authority. This includes:

- Service to the RICS;
- Service to another professional body;
- Market or industry recognition.

The Expert category relates to how a candidate has advanced, shared or interpreted knowledge in the following fields:

- Qualification;
- Teaching;

- Dispute resolution.

The Influencer category relates to how a candidate has influenced how a professional is perceived within the realms of:

- Leadership;
- Management;
- Development.

The Role Model category relates to how a candidate has exceeded standards for the benefit of clients, colleagues or the public in the following fields:

- Client care;
- Operation;
- Society and environment.

Candidates will need to select four relevant professional characteristics. There may be candidates who feel that more than four are relevant to them, however, only four need to be selected in support of their FRICS application. This means that candidates need to consider the supporting evidence available to them, allowing them to analyse which characteristics they are able to demonstrate most strongly.

In the Fellowship Applicant Guide (RICS, 2020m) (Table 9.1), the RICS provide a helpful table setting out the twelve professional characteristics in full. This includes details of the characteristic, a definition of the achievements required to meet the characteristic in question, the structure of the written statement and the third-party evidence required.

What Should the Personal Statement Include?

The personal statement is the candidate's opportunity to discuss their current and past roles, qualifications, academic accolades, and ambitions for the future. It should include sufficient detail to allow the assessment panel to understand how and why the candidate has met the standard required to become a Fellow. This should be over and above the standard required of a MRICS Chartered Surveyor.

Table 9.1 FRICS Professional Characteristics

Category	*Characteristic*	*Definition*	*Requirement*	*Written statement*	*Third party evidence*
Champion	1.Service to the RICS	Appointment to the RICS recognised group and performance of your role	A recognised group sits within the governance of the RICS or is required to make decisions on behalf of the RICS (for example, Professional Group Boards, Regional Boards, APC assessors).	State the group, your role and your performance in the role.	Confirmation from the chairperson or appropriate member of RICS staff or letter/citation confirming your appointment and performance of your role.
	2.Service to Another Professional Body	Fellowship or equivalent of another professional body	A professional body is typically an organisation with the primary aim to protect the public interest by the setting and regulation of standards for members of a profession. The equivalent of fellowship refers to a membership status that defines achievement beyond that expected by the majority of members of the professional body. The status must be awarded for achievement. Any professional body that gives an automatic right to fellowship after a certain time as an ordinary member, or only on payment of a higher fee, is not eligible.	State the award, the role of the professional body and the requirements to achieve the award.	Certificate or diploma.

(*Continued*)

Table 9.1 (Cont.)

Category	*Characteristic*	*Definition*	*Requirement*	*Written statement*	*Third party evidence*
			Note: Fellowship or equivalent grades of membership or professional qualifications that are recognised for APC enrolment or direct entry are not eligible. Other designations or certifications offered by industry associations or institutes may be eligible for the qualification characteristic.		
		Appointment to another professional body's recognised groups and performance of your role	A professional body is typically an organisation with the primary aim to protect the public interest by the setting and regulation of standards for members of a profession. A recognised group is one that sits within the governance of the body or requires decisions to be made on behalf of the body.	State the group, your role and your performance in the role.	Confirmation from the chairperson or appropriate member of the professional body's staff or letter/citation confirming your appointment and performance of your role.
	3.Market/ Industry Recognition	Appointment by a governance body	Appointment could be at local/regional/state/national/federal government levels or by multinational/international entities (for example, European Union, Association of South East Asian Nations, United Nations).	State the role you have been appointed to, the governance body and your responsibilities.	Confirmation from an appropriate member of the body's staff or letter/citation confirming your role.

	This could refer to meeting commitments to growth/reduction targets, introducing new legislation, improving standards in practice areas, developing new procedures/systems, or introducing value to products/services offered by RICS members. Your appointment is likely to be as an advisor, expert, or committee/panel member.		
Identified as an authority or leading figure by independent industry award	An award should follow a process of nomination, shortlisting and announcement of the winner. The award should showcase you as an important figure in your area of practice and/or location. The award must be credited to you, or work directly involving you, and not to your firm.	State the award you received, the nominated work, your involvement, and the process for gaining the award.	Award, certificate or letter/citation from the award organisers confirming the award for your work.
Identified as an authority or leading figure by media profile	You should be recognised and used by media to give opinions and advice for news items or provide regular commentary or features. Media could include television, radio, newspapers, magazines, websites, or social media.	State the type of contribution, the frequency and the topic.	Extract from contribution identifying your role or confirmation from an appropriate member of the media organisation's staff.

(*Continued*)

Table 9.1 (Cont.)

Category	*Characteristic*	*Definition*	*Requirement*	*Written statement*	*Third party evidence*
Expert	4.Qualification	Gained a higher education qualification	A higher education qualification is a qualification awarded by a learning institution that requires individuals to have previously completed degree-level or post-secondary study or possess appropriate career experience. For example: a postgraduate diploma/certificate, master's degree, or doctorate awarded by a university, college or business school.	State the qualification, the course structure, how it was delivered and the assessment.	Certificate or diploma.
		Gained an RICS certification	The RICS operate certifications that represent standards of competence in specialist and cross-sector areas. The certification must relate to improving standards and enforce additional quality assurance on members. Certifications required as a result of legislation or RICS Regulation are not eligible.	State the certification/ accreditation.	Not applicable. RICS database will be checked.
		Gained another professional qualification	These are qualifications that relate to a profession separate from the land, property, construction and infrastructure sectors but the skills of which positively influence the work of RICS members (for exam-	State the qualification and the awarding organisation.	Certificate or diploma.

5. Teaching	Published research or technical authorship	The publication should be professionally produced by a reputable independent organisation (print or online) – for example peer-reviewed professional journals, commercial textbooks or industry standards. Your involvement in writing should be as an authority on the subject. You do not need to be the sole or main author but could be a contributor.	State your involvement in writing the content, how it was developed and how and when it was published.	Extract from publication identifying your role or confirmation from an appropriate member of the publisher's staff.
	Innovator	This should be something new and creative that (for example) positively impacts the profession, improves current practice or introduces value to products/services offered by RICS members. It could refer to approaches to work, methods of working, software and hardware developments, or how firms operate within the industry.	State the innovation, your role in the development and the impact on practices.	Confirmation from your manager or an appropriate industry authority or a citation, report or article confirming your role.
	Academic, trainer or conference speaker	Employed or appointed by a reputable learning institute, training provider or conference producer to deliver your knowledge.	State your role, the organisation that employed your services and the knowledge you delivered.	Confirmation from an appropriate member of the organisation's staff.

(Continued)

Table 9.1 (Cont.)

Category	*Characteristic*	*Definition*	*Requirement*	*Written statement*	*Third party evidence*
	6.Dispute Resolution	Judicial appointment or recognition as a dispute resolver	Appointment by a third party to make determinations on disputes. Appointment can be as part of the judicial/legal system or by regulators, professional bodies or privately. Activities may include mediation, adjudication, expert witness, or arbitration or other defined dispute resolution methods.	State the appointment(s) and your role.	Confirmation from the third party that appointed you.
Influencer	7. Leadership	A leader operating at a senior level within an organisation and exercising extensive leadership and management skills	You must demonstrate your career progression and leadership and management of people and resources at a strategic level. If you own your business you must demonstrate leadership by the size and type of your contracts/clients.	State your role, its relation to the structure of the organisation and your responsibilities and impact as a leader, and describe your relationship with the person who is verifying your role.	Confirmation of your role from your manager, a colleague or other senior industry figure.

8.Manage-ment	A manager with direct responsi-bility for or influence over finances and people	You must be recognised as a key figure in your firm, by job title, job description, role on a board or impact of your recommendations, for financial management (of bud-gets, fee profiles or clients' money) and recruitment and performance of staff.	State your role, its relation to the structure of the organisation and your responsi-bilities and impact as a man-ager, and describe your relationship with the person who is verifying your role.	Confirmation of your role from your man-ager, a colleague or other senior industry figure.
9.Develop-ment	An individual with responsi-bility for or influence over a business	You must be recognised as a key figure in your firm, by job title, job description, role on a board or impact of your recommendations, for how your firm's services are communicated to stakeholders and how you generate and retain clients.	State your role, its relation to the structure of the organisation and your responsi-bilities for and impact on devel-oping your employer's busi-ness, and describe your relationship with the person who is verifying your role.	Confirmation of your role from your man-ager, a colleague or other senior industry figure.

(*Continued*)

Table 9.1 (Cont.)

Category	*Characteristic*	*Definition*	*Requirement*	*Written statement*	*Third party evidence*
Role model	10.Client Care	An individual who has developed an engaged reputation with clients or secured high profile engagements	You must demonstrate sustained client engagement and be recognised as a role model and key figure in your firm or be engaged on projects with a significant impact on the economy or society at an international, national, regional or local level. Significance could be determined by any media coverage, the clients or stakeholders.	State your role, your reputation with clients and colleagues and list your involvement on projects and their significance, and describe your relationship with the person who is verifying your role.	Confirmation of your role from your manager, a colleague or other senior industry figure, or a client testimonial.
	11. Operation	An individual who has developed or influenced business practices that improve how services are delivered	You must be recognised as a key figure in your firm, by job title, job description, role on a board or impact of your recommendations, for how your firm delivers services and maintains competitive advantage. It could refer to approaches to work, methods of working, software and hardware developments, technical training and development or recruitment of specialist staff.	State your role and your responsibilities for and impact on how your employer's business operates, and describe your relationship with the person who is verifying your role.	Confirmation of your role from your manager, a colleague or other senior industry figure.

12.Society and Environment	An individual who has contributed to charitable objectives, the improvement of living standards, or the sustainability of the environment	The dominant driver for doing the work should be improving or positively impacting economic development, society or the environment. It is not expected that work is voluntary or not-for-profit but your motivations should demonstrate commitment to social or environmental responsibility.	State your role, your responsibilities for and impact on improving society or the environment, and describe the objectives of the independent authority that is verifying your role	Confirmation of your role from an independent authority that contributes to improving society or the environment.

Candidates should consider their four professional characteristics when writing their personal statement. This could include, for example, discussing their recognition by an appropriate authority or discussing how they exceed standards for clients.

The personal statement is supported by the professional characteristics and written statements in support of each, which are discussed in further detail below. This section should total no more than 500 words.

What Should the Written Statements for the Professional Characteristics Include?

Four written statements are required for each professional characteristic, with no more than 500 words allocated to each. These will demonstrate how the candidate's achievements meet the requirements of the specific professional characteristic and reflect the Fellowship core principles discussed earlier on in this chapter.

The achievements or responsibilities detailed in each separate characteristic must be different, i.e., a candidate cannot use the same example for all four characteristics. This ensures that the candidate demonstrates a breadth and depth of experience and responsibilities which meet the core Fellowship principles.

Candidates should write the statements in as much detail as possible, clearly stating how they have met the required characteristic in the first person and past tense. For example, in relation to the Client Care characteristic, a candidate may state that 'I have demonstrated the client care characteristic by developing an engaged relationship and reputation with … and I have secured high profile engagements with …'.

Supporting third-party evidence should be detailed and attached for each professional characteristic selected. This should be a statement or copy document confirming the candidate's achievements. Candidates should submit as much relevant supporting documentation as possible, as this will enable the panel to make a confident decision that the candidate is of Fellowship standard.

If a candidate has submitted insufficient or inappropriate evidence, then the RICS may request further information or evidence which will delay the assessment of the candidate's submission. Alternatively, this may lead to an unsuccessful application, if the assessment panel simply do not feel that the candidate meets the requirements of the chosen professional characteristics.

The supporting evidence can be in a variety of forms. These include:

- Copy documents, e.g., certificates or awards;
- Press articles;
- Published articles by the candidate;
- Testimonials from third parties;
- Financial information;
- Website or written article confirming the candidate's achievements.

Third parties providing supporting evidence should not also be the candidate's referee. Any third party providing evidence may also be contacted by the RICS for verification of the candidate's achievements.

The assessment panel will use the following as a checklist when considering each professional characteristic, together with the candidate's supporting evidence:

- Does the achievement meet the stated definition and requirement of the characteristic?
- Does the third-party evidence validate the written statement?
- Is the characteristic demonstrated by an example different to the other characteristics?
- Is the achievement relevant to the RICS and the profession?
- Does the achievement demonstrate an intent to further the profession and the RICS?
- Does the achievement positively impact on a third party?
- Does the achievement promote the objectives of the RICS?
- Does the achievement reflect positively on the RICS?

All of these must be answered 'yes' for the candidate to be successful in any one professional characteristic.

Conclusion

Candidates achieving Fellowship standard are responsible for promoting the profession and the highest ethical standards. They are the current and future leaders of the profession and will be held in high regard by peers, clients and the wider public. This means that they must continue to act in the spirit of Fellowship, even after the letters, FRICS, have been awarded by the RICS.

FRICS is an extremely special designation to be bestowed on a Chartered Surveyor and the application process must not be undertaken lightly. That said, candidates should not be put off by the process, which is open to all MRICS surveyors, irrespective of age, gender and ethnicity.

The author of this book encourages all highly professional and ethical MRICS Chartered Surveyors to consider applying to become Fellows. This is because in 2018/2019, only 4% of Fellows (RICS, 2019b) were women. There is a clear need for increased diversity within the profession, as there are many highly skilled, professional and experienced surveyors working in practice who are not represented within the current Fellowship membership.

Hopefully this book goes some way to providing support to those wishing to be at the forefront of the surveying profession as a Fellow, providing clarity on the process, a role model for future FRICS members, and a confidence boost for those seeking Fellowship.

10 Registered Valuer Assessment

Introduction

In the previous chapters, we looked at the various RICS qualifications; AssocRICS, MRICS and FRICS. However, there are other qualifications, accreditations and certifications that Chartered Surveyors may wish to pursue depending on their discipline, sector and interests.

One of these is the RICS Valuer Registration Scheme (VRS). This is discussed here, in its own chapter, given its importance in maintaining the highest professional and ethical standards for valuation work.

In Chapter 11, we will look at a variety of other qualifications, accreditations and certifications which may be of interest to Chartered Surveyors.

What Is the VRS Scheme?

The VRS is operated by the RICS to monitor and regulate valuers. This relates to valuation work undertaken in accordance with RICS Valuation – Global Standards (Red Book Global) (RICS, 2020n). This is a document published by the RICS which sets out best practice standards for valuers.

The VRS scheme was introduced in 2011 in response to the financial crisis of 2007 onwards. It offers an alternative to the US system of appraiser licensing or valuer legislation in some European countries. Although it originated in the UK, it is now applied in a number of countries worldwide.

The aim of the VRS is to ensure the highest standards of quality assurance, ethics and consistency in valuation work. This is because

DOI: 10.1201/9781003156673-10

valuations are fundamental to good decision making by individuals and organisations, whether in relation to financial reporting, mortgage lending or portfolio acquisitions, amongst many other reasons. By promoting confidence in valuations, the VRS is helping to reduce risk in the industry and maintain a stable global economy.

This is emphasised by the following quote from a 2005 research paper; 'Property underpins a major proportion of financial decisions in mature economies. Failure to ensure assets are properly valued risks financial exposure for a wide range of stakeholders' (Gilbertson & Preston, 2005, p.1).

As a result, the VRS scheme is becoming increasingly recognised by the market and registered valuers continue to be in demand from both clients and employers.

What Is the Red Book Global?

The Red Book Global (RICS, 2020n) provides mandatory and best practice guidance for valuers. It was last updated in 2020 and is supported by national guidance, such as the UK National Supplement. In all cases, any national guidance must be read alongside the Red Book Global, rather than replacing it. This is important because it allows valuers to apply the Red Book Global to their national jurisdiction and ensures that the standards remain appropriate and applicable.

The Red Book Global applies to all written valuation advice, as well as oral valuation advice, to the fullest extent possible.

The Red Book Global includes two mandatory Professional Standards (PS):

- PS 1 – Compliance with standards where a written valuation is provided;
- PS 2 – Ethics, competency, objectivity and disclosures.

It also includes five Valuation Technical and Performance Standards (VPS):

- VPS 1 – Terms of engagement (scope of work);
- VPS 2 – Inspections, investigations and records;
- VPS 3 – Valuation reports;

- VPS 4 – Bases of value, assumptions and special assumptions;
- VPS 5 – Valuation approaches and methods.

The PS and VPS above are of mandatory application, unless one of five exceptions apply. In these instances, it may be inappropriate or impossible to apply the requirements of VPS 1–5.

As such, VPS 1–5 should be followed as far as possible, but with specific departures stated if required by a legislative or client requirement or in the context of a specific instruction, for example.

The five exceptions are set out in PS 1 of the Red Book Global:

- Providing valuation advice for agency or brokerage services, e.g., relating to the acquisition or disposal of asset(s). In these instances, valuers should refer to the RICS Professional Statement Real Estate Agency and Brokerage (3rd Edition). This could include advice on an offer to be proposed or which is being considered for acceptance. However, this does not encompass formal Purchase Reports which include a valuation and do not fall within this exception;
- Valuation advice provided in the role of an expert witness, as this advice will need to adhere to alternative guidance, e.g., RICS Practice Statement and Guidance Note Surveyors acting as expert witnesses (4th Edition);
- Valuations provided as part of a statutory function, such as taxation or compliance with specific legislation. To fall within this exception, the valuation needs to be provided in the performance of a statutory role with legislative powers exercised or enforced. It does not include valuations included in a statutory tax return (i.e., compliance only with the law), for example;
- Valuation advice provided for a client's internal purposes only, without liability and without disclosure to any third parties. This could include monthly portfolio valuations or advice on a proposed acquisition, for example;
- Valuation advice provided in the course of negotiations or litigation, for example in rent review or lease renewal negotiations.

There are also ten Valuation Practice Guidance Applications (VPGA), which relate to specific types of valuation instruction such as financial statements, secured lending and valuations of plant and equipment.

As such, valuers must join the VRS if they are providing any valuation advice which does not fall within the five exceptions outlined above.

Who Is Eligible to Register on the VRS?

VRS members are split into two categories: MRICS Registered Valuers and AssocRICS Registered Valuers. The difference should be noted, as some lenders may require a MRICS Registered Valuer to sign off valuation reports.

The RICS set out the requirements for being part of the VRS in the *Rules for the Registration of Schemes* (RICS, 2017a).

In Appendix 1, the Rules state that valuers may join the VRS if they have achieved the following:

- Valuation as a technical competency to level 3 in their APC assessment to become MRICS;
- Business valuation as a technical competency to level 3 in their APC assessment to become MRICS;
- Valuation as a technical competency to level 2 in their AssocRICS assessment.

Valuers may also join the scheme by the approved Direct Entry route, providing they also evidence three hours of CPD relating to the Red Book Global.

Some more experienced or mature valuers may have passed their APC prior to the current levels being introduced into the APC assessment. This could include members who have not undertaken the APC and became MRICS via the merger of the Incorporated Society of Valuers and Auctioneers (ISVA) with the RICS in 1999. In this case, a valuer will not need to undertake the VRS assessment but will be able to self-certify during their VRS registration that they have attained valuation competence to the required level. Any candidates in doubt should speak to the RICS about their current level of competence and how this aligns to the current APC levels.

Valuers must join the VRS if they are carrying out any Red Book compliant work, i.e., not relating to one of the aforementioned VPS 1–5 exceptions. This only applies in countries which have adopted the VRS as a mandatory requirement. The RICS provides a full current list of countries on its website.

How Does the Valuer Registration Process Work?

If a valuer meets the requirements to be registered under the VRS, they will need to undergo the process of valuer registration.

Valuer registration relates to the individual valuer, not RICS regulated firms. However, firms may sponsor (i.e., pay for) individual valuers under the VRS. If a valuer is being sponsored by their firm, then the firm will first need to apply to sponsor the valuer on the RICS website (via the online Regulation Portal) before the valuer can proceed with their own VRS registration.

There is a one-off registration fee to join the VRS and a yearly fee thereafter, which can either be paid by the firm or by the individual valuer. Firms who sponsor individual valuers will benefit from a discounted fee structure if multiple valuers are sponsored.

Valuers must register for the VRS online via the RICS website. The application process will require the valuer to provide information on the types of valuation that they deal with.

Registered valuers can work in both RICS regulated and non-regulated firms. However, in non-regulated firms, the valuer must ensure that a Complaints Handling Procedure is adopted with an RICS-approved redress mechanism and that adequate Professional Indemnity Insurance is held in accordance with the RICS Rules of Conduct.

Successful valuers who hold the RICS Registered Valuer designation will be listed in the RICS Directory, receive a VRS certificate and be able to use the RICS Valuer Registration designation and logo. For AssocRICS valuers, this must be stated as AssocRICS Registered Valuer, rather than just RICS Registered Valuer, and the VRS logo cannot be used.

How Does the VRS Assessment Process Work?

If candidates do not meet the requirements to register on the VRS outlined above, then they can undergo a separate VRS assessment to meet the eligibility requirements.

The VRS assessment requires candidates who are already MRICS to submit an application form, valuation-based experience, a case study and their CPD record.

As part of this, candidates need to identify what levels they need to demonstrate in terms of valuation competence. These accord to the current APC levels 1, 2 and 3. Valuation competence can be demonstrated by prospective MRICS VRS in either the Valuation or Valuation of Business and Intangible Assets competency. For Assoc-RICS candidates, there is only the option to pursue the Valuation competency. The Valuation of Businesses and Intangible Assets competency is not available.

For example, a candidate who achieved level 1 in their APC assessment will need to demonstrate levels 2 and 3 in their VRS assessment. A candidate who achieved level 2, will only need to demonstrate level 3. A candidate who achieved level 3 will not need to undergo the VRS assessment and can simply register on the VRS as discussed above.

It does not matter whether a candidate originally included Valuation as a technical competency in their APC assessment. In this instance, the candidate will need to demonstrate levels 1, 2 and 3 in their VRS assessment. This may apply if a candidate has changed roles or gained more experience in valuation over time. For example, in the case of a Building Surveyor or Quantity Surveyor who has become increasingly involved in HomeBuyer Reports with a valuation included.

The definition of each level is the same as those indicated in the APC assessment. These are:

- Level 1 – This relates to knowledge and understanding. This requires candidates to explain their knowledge and learning relevant to the description included in the competency guide. This could be through academic learning (e.g., a degree level course), CPD or on-the-job training. However, candidates should avoid too much repetition as their CPD record will be set out separately. Generally, level 1 will be met by including a brief list of knowledge or relevant topics, given the limitations of the overall word count, which will be discussed later in this chapter;
- Level 2 – This relates to the application of knowledge and understanding. Demonstrating level 2 relates to a candidate applying their level 1 knowledge and understanding in practice, i.e., through practical work experience. This relates to 'doing', rather than just knowing the theory and fundamentals. Candidates should refer to specific projects, instructions or examples to clearly demonstrate their practical activities and experience, including their role and relevance to the competency description;

- Level 3 – This relates to giving reasoned advice and demonstrating depth of knowledge. Candidates must expand on level 2 work experience or tasks to explain how they have provided reasoned advice to clients. This involves considering the options available and providing recommendations and advice on solutions. Candidates should ensure that they explain their role in the first person, using I, me and my, rather than relating to collective involvement using we, our and us.

Candidates should refer to the current RICS competency guidance for the Valuation or Valuation of Business and Intangible Assets competency to confirm what each level requires in terms of knowledge, activities and advice.

At level 3 this generally requires the candidate to have undertaken valuations for a range of purposes, property types and tenures, with reference to factors driving value, various service levels and to the Red Book Global and associated RICS valuation guidance.

Specifically, level 3 is defined by the RICS (RICS, 2018, p.79) to:

- 'Demonstrate practical competence in undertaking valuations, either of a range of properties or for a range of purposes;
- Demonstrate the application of a wide range of valuation methods and techniques;
- Be responsible for the preparation of formal valuation reports under proper supervision and provide reasoned advice;
- Demonstrate a thorough knowledge of the appropriate valuation standards and guidance and how they are applied in practice'.

In terms of the aggregate amount of work-based experience required under the VRS, a candidate needs:

- Fifty days to advance from level 2 to level 3, within twelve months of applying for the VRS assessment;
- Thirty days from level 1 to level 2, within twenty-four months of applying for the VRS assessment;
- Twenty days up to level 1, within twenty-four months of applying for the VRS assessment.

This means that the maximum requirement to progress to level 3 is one hundred days, i.e., adding together the requirements to progress to level 1, from level 1 to 2 and from level 2 to 3.

The candidate's work experience must be undertaken under the supervision of an existing RICS registered valuer, who must provide a written declaration to confirm this to the RICS. A candidate's work experience could include provision of valuation reports to clients and research for the preparation of valuation advice.

Alongside the supervisor declaration, the candidate must also submit the VRS application form. This requires the following to be completed:

- Personal details and competency confirmation, i.e., Valuation or Valuation of Businesses and Intangible Assets;
- Employer details;
- Professional experience, i.e., a brief CV and overview of scope and responsibilities relating to valuation experience;
- Candidate statement, outlining experience gained to level 3;
- Supervisor's declaration;
- Case study;
- CPD, relevant to valuation and the Red Book Global;
- Data protection;
- Candidate declaration.

The case study must be 1,000–1,500 words focussing on the candidate's competence at level 3. It must be based on three practical valuation examples, which required the candidate to be solely or primarily responsible for the valuation advice provided to the client. The examples must have been completed in the twelve months prior to submitting the VRS assessment.

The case study is split into five sections:

- Summary of each of the three valuations, including the asset valued, valuation date, client, purpose and legal system;
- Summary of the client's instruction and notable features for each valuation example;
- Process followed for each valuation example, including comparables, due diligence research, unusual factors or special considerations and any challenges that were overcome;

- Valuation advice given by the candidate in each example;
- Learning achievements, with a link to the required competency level.

There should be no appendices attached to the case study.

The word count for the valuation case study is limited, particularly as it requires candidates to include sufficient detail on three separate examples. Candidates should ensure that they write concisely, avoiding too much irrelevant detail or description. They should include the key details for each example and ensure that any defining or challenging aspects are emphasised. If any of the detail is common to more than one of the examples, then this may be combined to avoid substantial repetition and help to meet the word count requirements.

Effectively, the case study will then read similarly to an executive summary or an abstract to a valuation report, enabling the VRS assessors to understand the candidate's valuation advice and supporting justification for each example.

The RICS confirm that certain details of the three examples may be kept confidential if required by the client, e.g., names or locations. If any information is redacted, then a statement to this effect should be included in the case study.

In relation to the aggregate CPD requirements, the RICS require that:

- Ten hours are undertaken to move from level 2 to level 3;
- Ten hours are undertaken to move from level 1 to level 2;
- Twenty hours are undertaken to move up to level 1.

This means that the maximum CPD requirement is forty hours to progress from below level 1 up to level 3. A candidate's CPD should be structured with clear learning outcomes; examples include private study, organised learning or work-based learning.

For AssocRICS candidates wishing to become AssocRICS registered valuers, the above requirements of level 2 must be met, rather than the level 3 requirement for MRICS registered valuers.

AssocRICS candidates must, therefore, satisfy the requirements of level 2 only. This includes fifty days of total work experience split between thirty days from level 1 to level 2 and twenty days up to level

1. The CPD requirement is a total of thirty hours, split between ten hours from level 1 to level 2 and twenty hours up to level 1.

How Are VRS Assessments Considered by the RICS?

The VRS assessment is paper-based, with no in-person or online interview. It will be assessed by a current registered valuer and RICS trained assessor, who will assess the candidate's work experience against the required competency levels. This includes reviewing the case study content, structure and level of detail, including the standard of written language and professionalism. The case study is reflective of the quality of a valuer's reports to client, so poorly written or inaccurately proofread case studies will not be successful.

The VRS assessment result will be confirmed by the RICS within twenty working days. After this, successful valuers will need to undergo the process of valuer registration discussed earlier on.

If a candidate is not successful in their VRS assessment, a feedback report will be provided by the RICS and the candidate will be able to reapply for assessment after three months or more have passed.

Any elements which were successful can be resubmitted for up to one year. Any unsuccessful elements will need to be amended, along with an updated CPD record to meet the minimum RICS requirements. The referral report may also require the valuer to undertake additional work experience, up to a maximum of six months.

Candidates can appeal the decision within twenty-one working days of receiving the feedback report from the RICS. However, this can only be based on a failure in the process rather than a disagreement over the result.

How Does the VRS Operate in Practice?

The VRS focusses on the consistency and quality of the valuation process, rather than assessing whether valuation figures are correct. The RICS adopts a risk-based approach to monitoring valuers through the VRS.

The RICS initially create a risk profile based on the information submitted when a valuer applies for regulation under the VRS. RICS may require additional information to be provided by the valuer if a high risk rating is applied during this process. This risk profile is updated annually to reflect any changes to a valuer's workload.

The RICS also undertakes Desk Based Reviews (DBS) on a random basis. This may be where a valuer is identified as high risk by the annual risk profile or as a result of a complaint submitted to the RICS. As part of the DBS, the RICS will review the valuer and firm's terms of engagement, valuation reports and any other relevant documentation. The outcome of the DBS is provided in a written report which includes findings on compliance levels and advice on improvements.

In the case of poor compliance, the RICS may undertake a Regulatory Review Visit (RRV). Typically, this takes one to two days. This may be the result of a risk profile identifying high risk valuations in low numbers, high valuation volumes or evidence of poor valuation practice. The RRV can be in the form of a physical inspection at the valuer's office of files and processes or a remote inspection where the RICS requests and reviews relevant documents. Both will be followed up by a detailed discussion with a further feedback report provided to improve compliance.

There are no minimum requirements on VRS review frequency. However, if a large firm is subject to an RRV, then a minimum of 15% of the registered valuers will be sampled. The RICS state that 80% of RRVs have compliance ratings between satisfactory to very good (RICS, 2020o).

What Improvements Are Typically Identified in VRS Feedback Reports?

Examples of common areas for improvement following VRS feedback reports include:

- Ensuring compliant terms of engagement (VPS 1) and valuation reports (VPS 3) are issued to clients. Having standard templates on file can help to improve consistency and compliance;
- Undertaking and confirming robust conflict of interest checks, with a clear audit trail kept on file. Valuers should be familiar with the RICS Global Professional Statement Conflicts of Interest;
- Undertaking logical inspections as part of the valuation process, ideally following a standard proforma or checklist with a clear audit trail held on file;

- Ensuring comparable evidence is retained on file, with clear and detailed analysis and supporting verification. The final valuation figure must be justified and evidenced by the comparable evidence, which will help to defend any future negligence claims;
- Ensuring that valuation files are archived in good order, with an accompanying checklist to ensure that all relevant sections are included.

Conclusion

In conclusion, becoming a registered valuer will be a mandatory requirement for many Chartered Surveyors. This applies if they are undertaking any valuation work not subject to one of five exceptions under the RICS Red Book Global (RICS, 2020n).

Understanding the aims of the VRS is important, together with the processes and practices that should be adopted to ensure that high quality valuation advice is provided to clients.

In Chapter 11, we will look at other common qualifications, accreditations and certifications that Chartered Surveyors may be interested to pursue.

11 Other Qualifications

Introduction

In this chapter, we look at additional qualifications which Chartered Surveyors may wish to pursue. There are a wide variety of accreditations, further qualifications and certifications available, suiting a broad scope of sectors, specialisms and interests.

Surveyors seeking even more options and opportunities could also refer back to the discussion of Direct Entry requirements earlier in this book. These may provide interesting routes for career progression and diversification, as many of these memberships and accreditations have close links to the RICS and related surveying roles.

As well as the various professional qualifications on offer, Chartered Surveyors may decide to return to academia and complete a post-graduate degree or PhD in a relevant subject. They may also decide to become involved in academic research via an academic institution.

Another excellent opportunity for Chartered Surveyors is to consider training as an RICS APC or AssocRICS assessor, chair, Counsellor or mentor. These are all extremely rewarding roles and could potentially contribute towards a surveyor's FRICS application.

Pursuing any of these opportunities will count towards a surveyor's CPD requirements for the year. They will also help to further a surveyor's career prospects and allow them to expand into new or increasingly specialist lines of work. They will also keep a surveyor's knowledge up-to-date and ensure that a surveyor's career continues to be challenging, exciting and interesting.

The various qualifications have been categorised into areas of practice based on the APC pathways and separate specialist areas of

DOI: 10.1201/9781003156673-11

practice. However, surveyors' careers are varied and there is no reason why, for example, a Residential Surveyor could not seek further qualifications in Building Surveying, or a Commercial Property Surveyor could not seek further accreditation in sustainability-related issues.

This chapter does not provide an exhaustive or complete list, as there are numerous qualifications on offer within a wide range of fields, niches and sectors. This chapter also focusses on UK opportunities and there will be an even wider scope of opportunities offered globally. However, this chapter does provide a guide to some of the most popular qualifications which Chartered Surveyors may wish to pursue. It aims to provide a variety of ideas and areas for future research or discussion, depending on a surveyor's individual circumstances, aims and interests.

Building Surveying and Building Control

There any many further qualification opportunities for Building Surveyors and Building Control Surveyors.

The RICS run a Building Conservation Accreditation scheme relating to the protection and management of historic and heritage assets. This allows surveyors to lead grant-funded work for organisations such as Historic England and Historic Scotland. The assessment requires both a written submission and an interview. Surveyors with an interest in conservation may also become a Full Member of the Institute of Historic Building Conservation.

Building Surveyors may also decide to specialise in asbestos surveying, e.g., via the Royal Society for Public Health (RSPH) qualification.

Building Surveyors may also wish to pursue further skills and training in architectural technology, through becoming a Chartered Architectural Technologist (MCIAT). Equally, some surveyors may wish to become dual qualified as a RIBA-qualified Chartered Architect or Associate Member.

They may also wish to become a Member or Fellow of the Faculty of Party Wall Surveyors or join the Pyramus & Thisbe Club, if they specialise in providing party wall advice.

The Chartered Association of Building Engineers (CABE) is another popular option to pursue, via the MCABE qualification. Another alternative is provided by Membership of the Chartered

Institution of Building Services Engineers (CIBSE), whilst the Institute of Clerks of Works and Construction Inspectorate (ICWCI) offers membership for surveyors advising on and managing construction works through inspection.

Building Control Surveyors may benefit from CABE and CIOB qualifications, as discussed elsewhere in this chapter. Qualifications specific to Building Control are also provided by Local Authority Building Control (LABC). Individuals or companies can become Approved Inspectors in accordance with Section 49 of the Building Act 1984 and Regulations 3 and 5 of the Building (Approved Inspectors etc.) Regulations 2010, via approval by the Construction Industry Council Approved Inspectors Register (CICAIR).

Surveyors specialising in Building Surveying or Building Control may be interested in further qualifications relating to Project Management or Quantity Surveying and Construction.

Commercial Real Estate, Corporate Real Estate, and Property Finance and Investment

Commercial and Corporate Real Estate Surveyors may overlap disciplines, e.g., with Valuation, Facilities Management and Residential (e.g., mixed use properties).

There are also various niche areas of commercial and corporate professional practice where additional qualifications may be helpful.

For both public and private sector surveyors specialising in business rates, opportunities include becoming a Fellow or Corporate Member of the Institute of Revenues Rating and Valuation (IRRV) or joining the Rating Surveyors' Association. Similarly, surveyors advising on compulsory purchase instructions may benefit from joining the Compulsory Purchase Association (CPA).

There are also organisations which provide support and guidance to public and third sector surveyors, such as the Association of Chief Estates Surveyors and Property Managers in the Public Sector (ACES), Association of University Directors of Estates (AUDE) and the Charities' Property Association.

Various commercial property sectors and markets also have their own representative bodies, such as Accessible Retail (for retail warehouse and superstore property), British Property Federation, Property Managers Association (PMA), Revo (Retail Revolution) and the Shop

Agents Society. There are also organisations supporting a diverse and inclusive range of surveyors including Black, Asian and Minority Ethnicities (BAME) in Property, Freehold and the Association of Women in Property.

Surveyors specialising in providing advice on Property Finance and Investment often undertake further qualifications and training in financial modelling, such as those offered by Cambridge Finance or an academic institution. They may also be interested to become a member of the Investment Property Forum.

Land and Resources

The pathways related to Land and Resources have been categorised together as they cover a considerable range of roles and responsibilities held by Chartered Surveyors. These include Environmental Surveying, Geomatics, Infrastructure, Land and Resources, and Minerals and Waste Management.

For example, the RICS Property Measurer Certification requires surveyors to demonstrate consistent reporting in line with the International Property Measurement Standards (IPMS).

The Society for the Environment provides the Chartered Environmentalist qualification. This relates to the promotion of sustainable environments through policy making, best practice standards and professional advice. Surveyors working in this area of practice may also wish to become a Chartered Member of the Chartered Institute of Ecology and Environmental Management.

Additional opportunities are provided by the Chartered Institute of Environmental Health, which offers Chartered status for surveyors advising on environmental health issues, whilst the Energy Institute accredits surveyors advising on sustainability and energy efficiency-related issues.

Facilities Management

In Facilities Management (FM), an alternative professional body is the International Facilities Management Association (IFMA). Alongside the RICS, IFMA provides certificates in FMProfessional (FMP) and Sustainability Facility Professional (SFP) and certification as a Certified Facility Manager (CFM). The Institute of Workplace and

Facilities Management (IWFM) also provides membership options, e.g., Associate (AIWFM), Member (MIWFM), Certified (CIWFM) and Fellow (FIWFM).

The RICS Building Information Modelling (BIM) Certification may be of interest to surveyors who regularly use BIM to advise clients throughout the lifecycle of construction projects.

Management Consultancy

Surveyors specialising in management consultancy may decide to join another organisation such as the Chartered Management Institute (Member or MCMI, Chartered Member or CMgr MCMI or Chartered Fellow or CMgr FCMI) or the Chartered Institute of Personnel and Development (CIPD). They may also decide to pursue further post-graduate qualifications, such as a Master of Business Administration (MBA).

Planning and Development

Many Chartered Surveyors are also dual qualified as Chartered Town Planners (Member of the Royal Town Planning Institute, or MRTPI). This requires completion of one of the following routes to qualification; L-APC or DA-APC Licentiate, A-APC AssocRTPI or EP-APC Experienced Practitioner. There are also the Legal Associate (LARTPI) and Chartered Fellow (FRTPI) membership levels.

Project Management

Project Managers have a wide variety of alternative qualifications to pursue. All of these are valuable and will depend on the surveyor's specific role and sector. Some employers may prefer specific qualifications to be held, e.g., if they are particularly desirable to a key client.

The Chartered Institute of Building offers the Chartered Member (MCIOB) and Fellowship (FCIOB) qualifications. This recognises the highest standards of professionalism and technical skills in the construction industry.

Being PRINCE2 qualified is very popular for project managers to pursue. This provides a process-based method based on the concept of Projects IN Controlled Environments. Certifications include PRINCE2 Foundation and PRINCE2 Practitioner.

Project Managers may also wish to gain membership of the Project Management Institute (Project Management Professional, or PMP, for example) or Association for Project Management (Chartered Project Professional, or ChPP, for example).

Quantity Surveying and Construction

Similarly, to other disciplines, Quantity Surveyors may find that they decide to specialise in a certain niche sector, market or area of practice. Various qualifications detailed in the Building Surveying or Project Management categories may, therefore, be of interest.

Quantity Surveyors may also decide to become dual qualified, for example, as a Member or Fellow of the Chartered Institution of Civil Engineering Surveyors, Full Member of the Institution of Civil Engineers (ICE), Full Member of the Institution of Mechanical Engineers (IME), Institution of Structural Engineers (ISE) or Member or Fellow of the Association of Cost Engineers (ACostE).

Residential

Residential Surveyors typically advise on a wide range of instruction types, including building surveys, valuations, purchase/sale and leasing/lettings. Therefore, many of the other qualifications in this chapter may be of interest.

Residential Surveyors may be interested in joining the Residential Property Surveyors Association (RPSA) or the Independent Surveyors & Valuators Association (ISVA), which both represent independent surveyors. Additional residential-focussed qualifications are also offered by Sava, including Certificates in Residential Surveying and Residential Valuation.

Surveyors specialising in agency work may be interested in qualifications such as those offered within the Propertymark Suite by the National Association of Estate Agents (NAEA) and Association of Residential Lettings Agents (ARLA). Residential Property Managers may be interested similarly in becoming a Member or Fellow of the Institute of Residential Property Management (IRPM).

The Government are currently proposing mandatory licensing and qualifications to be held by estate and letting agents, which is likely to lead to an influx in candidates for the AssocRICS Real Estate Agency pathway. This will be a very positive step in increasing professional, ethical and technical standards in previously unregulated agency work.

Surveyors specialising in housing policy may wish to consider becoming a Member of the Chartered Institute of Housing (CIH).

Rural

There are three key organisations that are generally of interest to Rural Surveyors. The first is the Central Association of Agricultural Valuers (CAAV), of which the Fellow of CAAV (FAAV) qualification is often pursued at the same time as or shortly after becoming MRICS qualified. The second and third organisations are the Institute of Agricultural Management (IAgrM) and British Institute of Agricultural Consultants (BIAC).

Rural Surveyors involved in planning matters may also decide to become dual qualified as a Chartered Town Planner (MRTPI).

Valuation and Valuation of Businesses and Intangible Assets

The previous chapter took an in-depth look at the requirements of the VRS, which is the primary requirement for Valuation Surveyors.

However, other qualifications exist such as becoming certified in Entity and Intangible Valuations (CEIV) via the RICS.

Insolvency practitioners who advise on fixed charge or Law of Property Act 1925 receivership work may wish to pursue membership of the voluntary Registered Property Receivership Scheme. Membership of the Association of Property and Fixed Charge Receivers (NARA) may also be pursued at Fellow or Associate level.

Business and Intangible Asset Valuers may consider further qualifications provided by the Associated of Chartered Certified Accountants or decide to join the International Institute of Business Valuers (IIBV) or International Business Valuers Association (IBVA).

Dispute Resolution

Dispute resolution is an area of practice which relates to nearly all of the above surveying pathways. There are multiple accreditation schemes and certifications which Chartered Surveyors may wish to pursue relating to all areas of dispute resolution, e.g., acting as an expert witness or as a third party dispute resolver.

The RICS run a Dispute Resolution Service (DRS), which appoints third party dispute resolvers for a variety of different methods of Alternative Dispute Resolution. Surveyors can train to act in these roles through various RICS qualifications, e.g., Diploma in Arbitration or Adjudication and becoming an Accredited Mediator. There is also the Expert Witness Accreditation Scheme (EWAS), which relates to surveyors acting in the role of an expert witness during third party proceedings.

Other organisations also provide qualifications, such as the Chartered Institute of Arbitrators (MCIArb or FCIArb, a Chartered Arbitrator), Civil Mediation Council (CMC Registered Mediator), Expert Witness Institute (EWI) and International Mediation Institute (IMI).

Health and Safety

Health and Safety is another area of relevance to every single surveying pathway outlined above, and, indeed, every single Chartered Surveyor. There are a wide variety of qualifications on offer, including the Construction Skills Certification Scheme or membership of the Institute of Occupational Safety and Health (TechIOSH, CMIOSH or CFIOSH). In addition, surveyors working near to railways or trackside are likely to need a Personal Track Safety Card.

Surveyors may also wish to specialise in advising on fire safety or Fire Risk Assessments (FRA), e.g., becoming a member of the Institution of Fire Engineers (AIFireE, MIFireE or FIFireE) or holding the National Examination Board in Occupational Safety and Health (NEBOSH) Fire Safety Certificate.

The RICS has also recently launched a new training programme for Building Surveyors and Building Control Surveyors in response to the Building Safety Bill. This will train surveyors to carry out external wall system assessments for low to medium risk residential buildings.

Another relevant certification is being listed on the National Register of Access Consultants. This enables surveyors to advise on access requirements and improvements in line with the Equality Act 2010.

Conclusion

In conclusion, Chartered Surveyors have a wealth of opportunities to pursue after becoming MRICS or FRICS qualified. These will help to boost a surveyor's career prospects, as well as enabling them to provide the highest standard of professional advice in a focussed sector, market or area of practice.

In Chapter 12 we will look at how Chartered Surveyors can progress their career by setting up in practice as an RICS regulated firm.

12 Setting Up in Practice

Introduction

After qualifying as a Chartered Surveyor, some candidates will decide to pursue self-employment or to set up their own RICS regulated firm. This requires knowledge and application of many key RICS requirements, including holding adequate insurance, a CPD policy and a Complaints Handling Procedure (CHP). Setting up a new practice is not something to be undertaken lightly and requires robust knowledge of the legal and regulatory requirements, as well as having good business and financial sense.

We will not be looking at business planning or how to start or grow a business in this chapter. This is because there are many comprehensive resources already available which surveyors can refer to.

What Does Regulated by the RICS Mean?

In the first chapter of this book, we looked at how the RICS self-regulates firms. This provides clients and the public with confidence that a firm is acting in accordance with the highest standards of ethics, competence and professionalism.

How Can a Firm Register for Regulation by the RICS?

It is free to register a firm for regulation by the RICS. The process is simple and requires a Firm Details Form to be filled out and submitted to the RICS online.

DOI: 10.1201/9781003156673-12

The following details are required:

- Nomination of a Contact Officer – An RICS member who will be the main point of contact with the RICS;
- Nomination of a Responsible Principal – An RICS member who will ensure compliance with RICS regulation and standards;
- Firm name, contact details, type and registration numbers;
- Details of all RICS Principals (partners or directors) and employees.

After submitting the Firm Details Form, the firm will then need to provide further registration details using the RICS Firms Portal.

Firms wishing to register for regulation must meet the following requirements:

- The firm must offer professional surveying services to clients;
- The firm must have at least 25% of the Principals who are AssocRICS, MRICS or FRICS qualified. If a firm has over 50% principals who are AssocRICS, MRICS or FRICS then it must register for regulation as a mandatory requirement;
- The firm must comply with the RICS Rules of Conduct at all times.

Once a firm is regulated by the RICS, it can use the RICS logo and the designation 'Chartered Surveyors' within its trading name. There is a mandatory requirement to state that the firm is regulated by the RICS on all business material. The firm can also be listed in the RICS Find a Surveyor directory.

What Is Required of Firms by the RICS Rules of Conduct?

RICS regulated firms must follow the Rules of Conduct for Firms and all RICS members must follow the Rules of Conduct for Members. These will be discussed in further detail in Chapter 14.

The Rules of Conduct for Firms were last updated in March 2020 as Version 7. They are based on the principles of better regulation; proportionality, accountability, consistency, targeting and transparency. The Rules of Conduct are made under Article

18 of the Supplemental Charter 1973 and Bye-Law 5 of the RICS Bye-Laws.

The aim of the Rules of Conduct for Firms is to provide regulated firms with clear standards of professionalism, conduct and practice. A failure to comply with the Rules of Conduct may lead to disciplinary action by the RICS.

The Rules of Conduct for Firms are split into three parts; General, Conduct of Business, and Firm Administration.

Part I General requires firms to have a designated Contact Officer who will liaise between the firm and the RICS.

Part II Conduct of Business requires firms to:

- Act with integrity and avoid conflicts of interest;
- Carry out professional work competently;
- Provide a high standard of service to clients;
- Ensure procedures are in place for staff training and CPD;
- Operate a CHP and a complaints log, including an RICS-approved redress mechanism;
- Protect clients' money appropriately;
- Hold adequate Professional Indemnity Insurance (PII) cover.

Part III Firm Administration requires firms to:

- Promote their professional services truthfully and responsibly;
- Manage their finances appropriately;
- Have arrangements in place to cover the incapacity or death of a sole practitioner. This is typically by having a locum in place, e.g., an independent surveyor or another professional. They may also be appointed to be the firm's Complaints Handling Officer;
- Use the RICS designations and logo appropriately;
- Submit information to the RICS in a timely manner;
- Cooperate fully with the RICS.

The Rules of Conduct are currently being consulted on by the RICS to provide a new statement relating to Ethics and Rules of Conduct. This is likely to be published during 2021.

What Are the RICS Requirements Relating to Professional Indemnity Insurance?

RICS regulated firms must have Professional Indemnity Insurance (PII) cover in place, which meets the requirements of Rule 9 of the Rules of Conduct for Firms. PII protects firms from negligence claims made by clients who are not satisfied with the professional services provided to them. It may also cover a firm for loss of documents or data, intellectual property claims and defamation. PII protects clients from financial losses that a firm cannot meet.

The RICS require that PII must meet the following minimum requirements (RICS, 2020p):

- Provide cover for any one claim or be on an aggregate plus unlimited round the clock reinstatement basis;
- Fully retroactive ('claims made' basis);
- Include RICS' minimum policy wording. The latest changes were introduced to this in May 2020, including fire safety exclusion and External Wall System 1(EWS1) form requirements;
- Provide at least the minimum level of indemnity based on the firm's turnover in the previous year (or estimated for a new firm);
- Provide for a maximum level of uninsured excess based on the sum insured;
- Underwritten by a listed insurer;
- Provide cover for past and present employees.

What Are the RICS Requirements Relating to Handling Clients' Money?

Many regulated firms will handle clients' money in the course of their professional work. Any monies held must be handled appropriately via robust controls and systems.

The RICS provide requirements for client money handling in the Professional Statement Client Money Handling (1st Edition). Although not an exhaustive list, some of the key requirements include:

- Not holding office money in a client money account;
- Ensuring that the word 'client' is in the account name;

- Holding all monies in a client money account over which the firm has exclusive control;
- Ensuring that client money is immediately available;
- Confirming the bank operating conditions in writing.

Any firms holding client money must also pay an annual regulatory review fee following completion of their Annual Return. This contributes to operation of the RICS Client Money Protection Scheme, which provides recourse for clients when a regulated firm is unable to repay a client's money, subject to specific limits and exceptions.

What Are the RICS Requirements Relating to Complaints Handling Procedures?

All regulated firms must operate a Complaints Handling Procedure (CHP) in line with Rule 7 of the Rules of Conduct. Firms must also operate a complaints log, nominate a Complaints Handling Officer and include an approved redress mechanism within their CHP.

A firm's CHP should be in writing, in two stages; the first providing internal consideration of the complaint, and the second providing independent redress if the complaint cannot be resolved by the firm.

Clients must be made aware of the CHP in the firm's Terms of Engagement, as well as a copy of the CHP being provided to a client in the event of a complaint being received. If a complaint is received, the firm should also provide notice of this to their PII provider.

What Are the RICS Requirements Relating to Bribery, Corruption, Money Laundering and Terrorist Financing?

RICS regulated firms must comply with the mandatory RICS requirements relating to anti-bribery, corruption and money laundering. These are set out in the Professional Statement Countering Bribery and Corruption, Money Laundering and Terrorist Financing (1st Edition) and the 5th Anti Money Laundering Directive. This requires the firm to assess risk relating to money laundering and to put processes and procedures in place to carry out due diligence on clients, amongst other requirements.

What Are the RICS Requirements Relating to Continued Professional Development?

All RICS members (AssocRICS, MRICS and FRICS) must undertake a minimum of twenty hours of Continued Professional Development (CPD) each year. This must be completed by the 31st December and recorded online by the 31st January. At least ten hours must be formal (structured) CPD, and every three years CPD relating to ethics must be undertaken.

Formal CPD does not just include formal training courses, it includes anything with a clear learning objective or outcome. Examples include online training, webinars, structured discussions, technical authorship and undertaking academic courses.

Certain activities will not count as CPD, including networking, social events, committees and clubs, which have no clear link to a member's professional role.

Firms have a duty to ensure that staff training is structured and promotes provision of professional work to the required standards of skill, care and professionalism. This is likely to include a structured training programme, appraisal system and support for personal development. Firms will need to declare that this is in place as part of their Annual Return to the RICS.

What Other Statutory Requirements Do Firms Need to Comply With?

In addition to the above RICS requirements, there are a myriad of statutory requirements which a regulated firm may need to adhere to.

It should be noted that the below list is not exhaustive, but provides a starting point for any newly regulated firm:

- Business Names Act 1985 and Companies Act 1985;
- Health and Safety at Work Act 1974;
- Bribery Act 2010;
- 5th Anti Money Laundering Directive;
- Data Protection Act 2018 and the General Data Protection Regulation (GDPR);
- Employment law.

Firms also need to consider taxation requirements, e.g., VAT. Insurance, including public liability, employer's liability, building, contents, cyber and business interruption must also be considered.

Conclusion

In conclusion, setting up in practice is not something to be taken lightly by the newly qualified (or indeed experienced) Chartered Surveyor. This includes ensuring that a firm is correctly registered for regulation by the RICS and complying with the requirements of the RICS and any additional, statutory and business-related requirements. Chapter 14 will look further at the RICS ethical and professional requirements expected from both firms and members.

13 Global Perspectives

Introduction

So far, we have primarily taken a UK-centric view of how to become a Chartered Surveyor. However, Chartered Surveyors have global reach and operate in the majority of countries worldwide.

Many surveyors will be employed directly by national companies in their home country, whereas others may be employed by multinational organisations headquartered in another country. There are various tax, legal and financial implications to consider, particularly if a surveyor relocates to work in another country. Specialist advice should be sought in all circumstances, which is outside the scope of this book.

In this chapter, we will look at how Chartered Surveyors are viewed globally, with a vast number of challenges and requirements being highlighted.

Does the RICS Operate Globally?

The RICS is a global governing body, with world regions split into the Americas, Europe, Asia Pacific and Middle East and Africa. The RICS (RICS, 2020a) reports that there are approximately 130,400 qualified and trainee professionals worldwide. Approximately 3,300 are located in the Americas, 107,000 in the Europe, Middle East and Africa (EMEA) region and 20,100 in the Asia Pacific region.

There are other national governing bodies and associations, together with regional and national legislative requirements, that surveyors will need to be aware of. Most importantly, surveyors need to be aware of anything that affects the professional advice that they

DOI: 10.1201/9781003156673-13

give to their clients. For example, Landlord and Tenant specialists working in Scotland will need to be aware that the English Landlord and Tenant Act 1954 does not apply. Similarly, in Australia surveyors will need to comply with health and safety legislation contained within the Work Health and Safety Act 2011 (WHS Act).

It is recommended that anyone looking to become a Chartered Surveyor should check the requirements of the RICS, government or other bodies to work as a surveyor in their country of choice. In some countries, this may require mandatory regulation, licensing or specific qualifications.

The requirement for surveyors to hold an RICS qualification has become more important as surveying and consulting firms have become increasingly globalised. This means that the MRICS and FRICS Chartered Surveyor membership levels are widely recognised as a gold standard of professionalism. The AssocRICS level of membership is also becoming increasingly recognised globally, although perhaps at a slower pace than MRICS and FRICS.

Being a Chartered Surveyor is likely to open up opportunities to work abroad in a variety of different industries, sectors and environments. There may also be opportunities for surveyors to work on foreign projects on an interim, short or long-term basis.

Many other professional bodies have APC or AssocRICS Direct Entry arrangements with the RICS, as detailed in an earlier chapter. This means that surveyors registered or regulated by other national or regional bodies may be eligible to qualify as AssocRICS or MRICS either without undergoing any assessment or for the APC via the Preliminary Review route.

Examples of eligibility for APC Direct Entry without undergoing the final assessment interview or Preliminary Review route include:

- Australia – Associate Member or Member and Certified Quantity Surveyor of the Australian Institute of Quantity Surveyors (AIQS) with five years of post-qualification experience;
- Germany – OPA – Oberprüfungsamt des Bundes – Vermessungs-assessor, with completion of the second German state examination;
- Kenya – Being a Corporate Member or Fellow of the Institute of Quantity Surveyors of Kenya (IQSK) with five years of relevant post-qualification experience;
- South Africa – Registered Professional Valuer (RPV) of the South African Council for the Property Valuers Profession (SACPVP);

- USA – Senior Grade Member (MAI) of the Appraisal Institute (AI) with ten years of relevant experience.

Other global qualifications provide eligibility for APC Direct Entry via the Preliminary Review route, Senior Professional or Specialist assessments. Examples include:

- China – Full Member of the Shanghai Construction Cost Association (SCCA);
- Germany – HypZert (F) Valuers;
- USA – Full Member of the Society of Industrial and Office Realtors (SIOR).

Examples of global qualifications allowing for AssocRICS Direct Entry include:

- Australia – Level 1 or 2 Licensed Building Certifier or Affiliate or Associate Member of the Australian Institute of Quantity Surveyors (AIQS);
- Hong Kong – Government Survey Officer (Engineering);
- Saudi Arabia – Accredited Members (Real Estate division) of the Saudi Authority for Accredited Valuers (TAQEEM);
- Trinidad and Tobago – Full Member of the Institute of Surveyors of Trinidad and Tobago (ISTT).

UK-qualified Chartered Surveyors looking to work abroad should become familiar with local, regional and national language, culture and market. This is essential to providing the highest standard of professional service, in addition to ensuring that the highest ethical standards are maintained. Cultural sensitivities and ways of doing business differ across the globe, so Chartered Surveyors need to be aware of these to build and maintain good client and peer relationships.

Chartered Surveyors in Europe

The RICS opened its Brussels office in 1993, with over 8,454 AssocRICS, MRICS or FRICS surveyors being members of the RICS in 2019 (RICS, 2019c). Of these, 1,273 were Registered Valuers. The split between countries is shown in Figure 13.1.

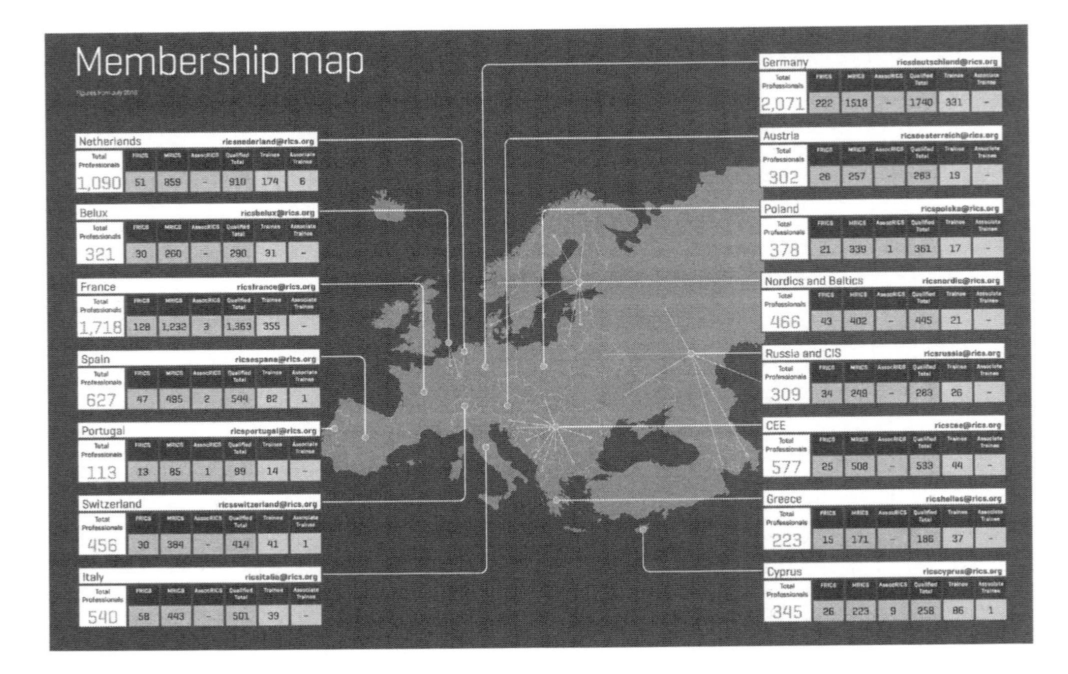

Figure 13.1 RICS European Membership Map

In Ireland, the Society of Chartered Surveyors Ireland (SCSI) provides dual membership of the RICS. Members of the SCSI (MSCSI) must complete the APC and will also be dual MRICS qualified. Similarly, MRICS Chartered Surveyors can apply to become MSCSI if they undertake work primarily in Ireland.

In Germany, there are a number of professional organisations, alongside RICS, which represent surveyors. For valuation surveyors, these include HypZert GmbH and the European Association of Certified Valuators and Analysts (EACVA). The RICS VRS also applies in Germany for any surveyors carrying out Red Book valuation work.

The RICS must recognise other relevant EU professional qualifications under the EU Directive Assessment, 2005/36/EC. This means that the RICS must consider practising surveyors' qualifications from other EU countries, providing that the individual surveyor's qualifications and experience meet the requirements of RICS membership. The approval process takes around four months, and applications are considered on a case-by-case basis. At the time of writing, the RICS had not provided any update on the application of the Directive post-Brexit in 2021 and beyond.

In Europe, there is the potential for language to be a barrier when working as a Chartered Surveyor in non-English speaking countries. In large UK or USA based firms, English may be the primary language spoken although knowledge of the local language, culture and market will be essential for Chartered Surveyors to provide the highest standards of client advice and service. Chartered Surveyors considering working in non-English speaking countries would be well advised to learn the language early on.

Chartered Surveyors in the Middle East and North Africa

The RICS has become increasingly active in the Middle East and North Africa (MENA) region, particularly as large firms have expanded their operations here. There is an increasing acceptance and preference for RICS qualifications, with a Market Advisory Panel and national and local associations operating in UAE, Qatar, Bahrain, Oman, Lebanon and Saudi Arabia. There is also a MENA branch of the RICS Alternative Dispute Resolution (ADR) Presidents Panel, which appoints third party dispute resolvers and expert witnesses.

Dubai provides a particularly interesting example of state regulation of the valuation profession, following the global recession in 2007–2008. Relevant bodies include the Dubai Land Department (DLD), Dubai Real Estate Appraisal Centre and the Real Estate Regulatory Agency (RERA).

The DLD requires UAE nationals to have two years of valuation experience, which is increased to five years of experience for non-UAE nationals (with at least six months of this being within the UAE). Valuation surveyors must initially pass the Dubai Real Estate Institute's Valuer Orientation Course and, as an annual regulatory requirement, valuers must submit three valuation reports each year to the DLD.

The RICS has no reciprocal arrangement with the DLD, so separate registration with each is required. Valuers must comply with the Emirates Book Valuation Standards (EBVS), which reflects the International Valuation Standards (IVS) and Red Book Global.

Chartered Surveyors in the Americas

The RICS in the Americas comprises twenty chapters, with primary offices based in New York and Toronto.

In the USA, there are other organisations providing guidance and membership to surveyors, including the American Society of Appraisers and the Appraisal Institute. Both are members of The Appraisal Foundation (TAF), which sets standards authorised by US Congress, and qualifications for valuers, or real estate appraisers as they are known in the USA.

Valuers in the USA must be licensed and certified under the provisions of the Financial Institutions Reform, Recovery and Enforcement Act 1989, which are implemented by individual states. Valuation work must be undertaken in accordance with either IVS and the TAF Uniform Standards of Professional Appraisal Practice (USPAP) or the Red Book Global.

In Canada, there are a number of organisations, including the Appraisal Institute of Canada, Canadian Association of Consulting Quantity Surveyors (CACQS) and the Canadian Institute of Quantity Surveyors (CIQS).

Chartered Surveyors in Africa

A wide number of African countries regulate surveyors via national organisations, such as:

- Kenya – Institution of Surveyors of Kenya (ISK) and Institute of Quantity Surveyors in Kenya (IQSK);
- Nigeria – Nigerian Institution of Estate Surveyors and Valuers and the Nigerian Institute of Quantity Surveyors (NIQS);
- South Africa – South African Institute of Valuers, Office of the Valuer General and the Association of South African Quantity Surveyors (ASAQS). Statutory registration is also compulsory;
- Uganda – Institution of Surveyors of Uganda (ISU).

Many surveyors in Africa dual qualify with a national association and as MRICS or FRICS.

Chartered Surveyors in Australasia

The RICS is active in Australia and New Zealand, with offices in Sydney, Auckland and Brisbane.

Australian surveyors may also register with the Australian Property Institute (API) or the Australian Institute of Quantity Surveyors (AIQS).

In New Zealand, there are two primary surveying organisations, the Property Institute of New Zealand (PINZ) and the New Zealand Institute of Quantity Surveyors (NZIQS). Valuers in New Zealand are regulated by the Valuers Registration Board (VRB) and the Valuers Act 1948. The key requirements for being regulated are having a valuation degree from an approved New Zealand university or being qualified with an approved professional body (e.g. MRICS or FRICS), together with having at least three years of relevant experience. Valuers must renew their regulation certificate annually.

Valuers in New Zealand must adhere to the Australia and New Zealand Valuation and Property Standards (PINZA Standards) or the Red Book Global and IVS.

What Other Global Organisations Provide Guidance to Chartered Surveyors?

There are a number of international bodies providing global guidance to Chartered Surveyors and other built environment professionals.

This chapter will look at five of these;

- International Valuation Standards (IVS);

- International Property Measurement Standards (IPMS);
- International Fire Safety Standards (IFSS);
- International Construction Measurement Standards (ICMS);
- International Land Measurement Standards (ILMS).

The IVS are set by the International Valuation Standards Council (IVSC). They aim to provide global consistency, transparency and confidence in valuation, with the valuation principles adopted by over one hundred and forty member organisations in over one hundred and fifty countries. For example, the RICS incorporates the principles of IVS into RICS Valuation – Global Standards (Red Book Global) (RICS, 2020n). The RICS also publishes national supplements and jurisdiction guides, e.g., the UK National Supplement, to contextualise the application of the Red Book in individual countries.

Furthermore, in Europe the European Group of Valuers' Associations (TEGoVA) publishes the European Valuation Standards (EVS). The EVS are supported by seventy-two professional bodies in thirty-eight European countries, including the RICS.

IPMS are set by the International Property Measurement Standards Coalition (IPMSC), including over eighty member organisations. IPMS aims to increase global consistency in measurement, with standards for offices, residential, retail and industrial now published. IPMS are then adopted by member organisations. For example, in RICS Property Measurement (2nd Edition), IPMS Residential and Offices have been adopted. The other standards are not yet adopted in the UK, although it is likely that they will be in the next edition of RICS measurement guidance.

IFSS are global fire safety standards, introduced in response to the 2017 Grenfell Tower fire in the UK. The aim is to provide international guidance throughout the building lifecycle that will ensure the safety of building design, construction, maintenance and occupation.

ICMS were introduced in 2017 to provide global consistency in the reporting, grouping and classifying of construction costs. ICMS are supported by forty member organisations, with the standards set by twenty-seven experts from seventeen countries. The 2nd Edition of ICMS now includes reference to life cycle costs and provides guidance on comparisons to the New Rules of Measurement (NRM).

Finally, ILMS aim to provide global consistency in the measurement of and due diligence relating to land, particularly for acquisition and transfer purposes. Alongside the UN Sustainable Development Goals, these aim to support good land administration particularly in developing countries. ILMS were introduced in 2016 and are now supported by over thirty member organisations.

Conclusion

It is clear from this chapter that Chartered Surveyors operate globally, with some countries having their own national organisations or legislative requirements. It is of paramount importance for Chartered Surveyors to be aware of relevant legislation, regulation and guidance in the country of their practice.

In Chapter 14, we look at ethics and professionalism, which should be held in high regard by all Chartered Surveyors, irrespective of where they work across the globe.

14 Professional Ethics

Introduction

Acting professionally and ethically is at the heart of what it means to be a Chartered Surveyor. In this chapter, we will look at the requirements of the RICS in relation to ethical and professional practice.

Most surveyors will initially be introduced to ethics and professionalism in a module at university (or another academic institution). This will then be built on by any workplace induction, ongoing training and CPD that a surveyor undertakes. Ethics is an area for continued focus throughout a surveyor's career, not just something that should be considered once at the outset.

Ethics, Conduct Rules and Professionalism is a mandatory APC competency that all candidates must satisfy at level 3. They must consider the ethics of their reasoned advice to clients and ensure that they give advice in a robust, professional and moral manner. This is not simply about knowing how to act ethically or being aware of the relevant RICS guidance; it is about the reasoned application of ethics to the decisions and actions that a surveyor makes within their personal and professional lives. In the final assessment interview, an incorrect or unethical answer in relation to the Ethics, Conduct Rules and Professionalism competency will constitute an automatic fail, irrespective of a candidate's performance in the rest of their interview.

AssocRICS candidates must also demonstrate their strong understanding and application of ethics and professionalism through the RICS online ethics module and test.

DOI: 10.1201/9781003156673-14

Once qualified as AssocRICS or MRICS, surveyors must undertake CPD in relation to the Global Professional and Ethical Standards every three years to satisfy the mandatory requirements of the RICS.

Acting ethically leads to high standards of professionalism, service and client care being provided, so read on to find out what surveyors need to know in practice.

Why Is Acting Ethically and Professionally so Important?

Acting ethically and professionally gives clients and the public confidence in the advice that Chartered Surveyors provide. This is emphasised by RICS research which stated that 'high ethical standards should be seen as part of an employer organisation's good governance and through this, its competitive advantage, because offering a high quality service to the public will raise the profile of the employing organisation (which is likely to be reflected in its balance sheet) and thus enhance the reputation of the entire surveying profession as well as that of its employees' (Plimmer, et al., 2009, p.5).

What Ethical Standards Should Chartered Surveyors be Aware of?

The RICS set ethical requirements for Chartered Surveyors in the Rules of Conduct for Members and the Global Professional and Ethical Standards. Firms also need to adhere to the Rules of Conduct for Firms which were discussed at length in Chapter 12.

The RICS also publish supporting guidance, including the Professional Statements Client Money Handling (1st Edition), Conflicts of Interest (1st Edition) and Countering Bribery and Corruption, Money Laundering and Terrorist Financing (1st Edition). There is also supplemental specific guidance in Professional Statement Conflicts of Interest – UK Commercial Property Market Investment Agency (1st Edition).

Chartered Surveyors will also need to adhere to sector specific standards set out by the RICS, such as RICS Valuation – Global Standards (Red Book Global) and the Black Book Suite of Guidance Notes for Quantity Surveying and Construction professionals.

Chartered Surveyors will also need to adhere to their own personal ethical and moral values, as well as those set out in their firm or employer's policy and guidance.

What Are the Rules of Conduct for Members?

The Rules of Conduct for Members reflect the requirements of the Rules of Conduct for Firms. They were last updated in March 2020 as Version 7. The Rules of Conduct for Members adopt the principles of better regulation; proportionality, accountability, consistency, targeting and transparency.

The Rules of Conduct are made by the RICS under Article 18 of the Supplemental Charter 1973 and Bye-Law 5 of the RICS Bye-Laws.

The aim of the Rules of Conduct for Members is to provide members with clear standards of professionalism, conduct and practice. A failure to comply with the Rules of Conduct may lead to disciplinary action by the RICS.

The Rules of Conduct for Members are split into two parts; General (which is very brief) and Personal and Professional Standards.

Part II Personal and Professional Standards requires members to:

- Act with integrity and avoid conflicts of interest;
- Carry out professional work competently;
- Provide a high standard of service to clients;
- Comply with the RICS requirements relating to CPD;
- Ensure that personal and professional finances are managed diligently;
- Submit information to and cooperate with the RICS as required.

What Are the Global Professional and Ethical Standards?

The RICS Global Professional and Ethical Standards were introduced in 2009 to replace the former twelve Core Values.

They reflect the requirements of the International Ethics Standards (IES), which were introduced in December 2016 and are adopted by over one hundred and twenty organisations globally.

The IES include the following ten principles: accountability, confidentiality, conflict of interest, financial responsibility, integrity, lawfulness, reflection, standard of service, transparency, and trust.

The five RICS Global Professional and Ethical Standards are to:

- Act with integrity – This relates to acting honestly and in a straightforward manner. Examples of this include respecting client confidentiality, upholding a surveyor's professional duty of care towards a client, avoiding conflicts of interest, not accepting gifts or hospitality that might be perceived as improper and providing unbiased professional advice to clients;
- Always provide a high standard of service – This relates to providing the highest standard of professional advice or support to clients in compliance with signed written Terms of Engagement. Examples of this include acting within a surveyor's scope of competence, being transparent about fees, communicating effectively and ensuring good payment practices for suppliers or contractors;
- Act in a way that promotes trust in the profession – This relates to acting professionally and positively in surveyors' personal and professional lives. Examples of this include fulfilling a surveyor's obligations to clients and other key stakeholders, promoting the benefits of acting professionally and acting ethically in all areas of life;
- Treat others with respect – This relates to acting ethically, fairly and in a non-discriminatory manner towards others no matter who they are or what they stand for. This will include discrimination based on any of the protected characteristics under the Equality Act 2010 such as age, sexual orientation or religion. Examples of this include treating all potential and existing clients fairly, equally and politely;
- Take responsibility – This relates to being accountable and taking action if something doesn't feel right. Examples of this include acting diligently, resolving complaints and taking a stand and speaking out where necessary.

Chartered Surveyors may benefit from referring to the RICS ethics decision tree (RICS, 2020q) when considering how to proceed in relation to an ethical dilemma.

The Rules of Conduct for Firms and for Members and the Global Professional and Ethical Standards are currently being consulted on by the RICS to provide a new statement relating to Ethics and Rules of Conduct. This is likely to be released during 2021 with the aim of providing clarity, effectiveness and brevity in refined ethical and professional standards.

How Can Chartered Surveyors Manage Conflicts of Interest?

To act ethically and professionally, Chartered Surveyors need to ensure that they are not influenced by any perceived or actual conflicts of interest.

The RICS set out mandatory requirements relating to conflicts of interest in the Professional Statement Conflicts of Interest (1st Edition). This relates to both the actions of firms and members and provides guidance on how to comply with the requirements of the Rules of Conduct and Global Professional and Ethical Standards.

The RICS state that 'the most important reason for avoiding conflicts of interest is to prevent anything getting in the way of your duty to advise and represent each client objectively and independently, without regard to the consequences to another client, any third party, or your own interests and that the clients and in turn the public can be confident you are doing so' (RICS, 2017a, p.6).

A conflict of interest is defined as (RICS, 2017a, p.4):

- 'Party conflict – a situation in which the duty of an RICS member (working independently or within a non-regulated firm or within a regulated firm) or a regulated firm to act in the interests of a client or other party in a professional assignment conflicts with a duty owed to another client or party in relation to the same or a related professional assignment;
- Own interest conflict – a situation in which the duty of an RICS member (working independently or within a non-regulated firm or within a regulated firm) or a regulated firm to act in the interests of a client in a professional assignment conflicts with the interests of that same RICS member/firm (or in the case of a regulated firm, the interests of any of the individuals within that regulated firm who are involved directly or indirectly in that or any related professional assignment);

- Confidential information conflict – a conflict between the duty of an RICS member (working independently or within a non-regulated firm or within a regulated firm) to provide material information to one client, and the duty of that RICS member (working independently or within a non-regulated firm) or of a regulated firm to another client to keep that same information confidential'.

The RICS require that:

- Regulated firms and members must not accept an instruction where there is an actual or perceived conflict of interest, unless written informed consent is provided by all concerned parties;
- Regulated firms must implement procedures to manage conflicts of interest, including written records of decisions made and controls put in place.

In the first instance, avoiding conflicts of interest is the preferred course of action. If a conflict of interest is identified, then the firm or member should consider whether the instruction should simply be declined or if the conflict can be managed via seeking informed written consent and/or implementing an information barrier.

An information barrier, previously known as a Chinese Wall, requires the 'physical and/or electronic separation of individuals (or groups of individuals) within the same firm which prevents confidential information passing between them' (RICS, 2017, p.4). This could include separating administration teams and filing systems or handling the instructions in separate offices and by different surveyors. Particularly in small firms, it may be impossible to set up a truly robust information barrier, so declining an instruction may be the most ethical course of action.

The RICS also provide supplementary mandatory guidance for UK commercial investment agency surveyors through the Professional Statement Conflicts of Interest – UK Commercial Property Market Investment Agency (1st Edition). This specifically prohibits the practice of dual agency (also known as double dipping), meaning that an agent cannot act for both the seller and buyer (i.e., both parties) in an investment transaction. The Professional Statement also provides guidance on multiple introductions and incremental advice.

Conclusion

In conclusion, it is clear that acting ethically is at the absolute heart of what it means to be a Chartered Surveyor. It is not simply enough to be sufficiently competent, skilled and knowledgeable. This is clearly evidenced by the requirement for all APC candidates to satisfy the requirements of level 3 in the Ethics, Rules of Conduct and Professionalism competency. In the final assessment interview, an incorrect or unethical answer in relation to this competency will constitute an automatic failure, irrespective of a candidate's performance in the rest of their interview. AssocRICS candidates must also demonstrate their strong understanding and application of ethics and professionalism through the RICS online ethics module and test.

Chartered Surveyors must seek to display the highest standards of professionalism in all areas of their lives, both personal and professional. This should be an aspiration for all existing and future Chartered Surveyors, irrespective of sector, market or discipline. It is a mandatory, not a voluntary requirement, and never something to be taken lightly.

There will be times when a surveyor will need to take responsibility to rectify a problem, challenge or mistake and it is the desire and requirement to act ethically that will ensure that these situations are dealt with diligently and professionally. Undertaking the mandatory three yearly CPD relating to the Global Professional and Ethical Standards once qualified as AssocRICS or MRICS will support surveyors in making ethical, professional and moral decisions in times of uncertainty or difficulty.

References

Gilbertson, B. and Preston, D., 2005. A Vision for Valuation. *Journal of Property Valuation & Investment*, 23 (2), pp. 123–140.

Gov.uk, 2020. Introduction of T Levels. [Online] Available at: https://www.gov.uk/government/publications/introduction-of-t-levels/introduction-of-t-levels [Accessed 12 October 2020].

Plimmer, F., Edwards, G. and Pottinger, G., 2009. *Ethics for Surveyors: An Educational Dimension. Commercial Real Estate Practice and Professional Ethics*, London: RICS.

RICS and Macdonald & Company, 2019. UK Rewards & Attitudes Survey 2019. [Online] Available at: https://www.macdonaldandcompany.com/rewards-attitudes-survey [Accessed 2019].

RICS, 2017a. Professional Statement Conflicts of Interest 1st Edition, London: RICS.

RICS, 2017b. Rules for the Registration of Schemes, London: RICS.

RICS, 2018. RICS Requirements and Competencies Guide, London: RICS.

RICS, 2019a. Counsellor Guide, London: RICS.

RICS, 2019b. More Female Role Models Can and Should Become FRICS. [Online] Available at: https://www.rics.org/uk/surveying-profession/career-progression/become-an-rics-fellow-frics/more-female-role-models-can-and-should-become-frics/ [Accessed 15 October 2020].

RICS, 2019c. Europe Digest 2019, London: RICS.

RICS, 2020a. Benefits of a RICS Qualification. [Online] Available at: https://www.rics.org/uk/surveying-profession/join-rics/benefits-of-rics-membership/ [Accessed 12 October 2020].

RICS, 2020b. Gender Statistics, London: RICS.

RICS, 2020c. Royal Charter. [Online] Available at: https://www.rics.org/globalassets/rics-website/media/governance/royal-charter/ [Accessed 5 October 2020].

RICS, 2020d. RICS Bye-Laws. [Online] Available at: https://www.rics.org/globalassets/rics-website/media/governance/bye-laws/ [Accessed 5 October 2020].

RICS, 2020e. RICS Regulations. [Online] Available at: https://www.rics.org/globalassets/rics-website/media/upholding-professional-standards/regulation/regulations/ [Accessed 5 October 2020].

RICS, 2020f. Governance Procedures and Processes. [Online] Available at: https://www.rics.org/globalassets/rics-website/media/governance/standing-orders/ [Accessed 5 October 2020].

RICS, 2020g. RICS Courses. [Online] Available at: http://www.ricscourses.org/Course/ [Accessed 12 October 2020].

RICS, 2020h. APC Candidate Training Plan. [Online] Available at: https://www.google.com/url?sa=t&rct=j&q=&esrc=s&source=web&cd=&cad=rja&uact=8&ved=2ahUKEwjOjJ3tn6_sAhUkSBUIHXfeAKAQFjAAegQIBxAC&url=https%3A%2F%2Fwww.rics.org%2Fglobalassets%2Frics-website%2Fmedia%2Fassessment%2Frics-apc-candidate-training-plan.doc&usg=AO [Accessed 12 October 2020].

RICS, 2020i. Annex A: Examples of Types Formal and Informal CPD Activities. [Online] Available at: https://www.rics.org/globalassets/rics-website/media/upholding-professional-standards/regulation/media/cpd-annex-a-160518-mb.pdf [Accessed 16 October 2020].

RICS, 2020j. APC Candidate Guide, London: RICS.

RICS, 2020k. Senior Professional Assessment Applicant Guide, London: RICS.

RICS, 2020l. Specialist Assessment Applicant Guide, London: RICS.

RICS, 2020m. Fellowship Applicant Guide. [Online] Available at: https://www.rics.org/globalassets/rics-website/media/qualify/fellowship-applicant-guide-rics.pdf [Accessed 15 October 2020].

RICS, 2020n. RICS Valuation – Global Standards. [Online] Available at: https://www.rics.org/uk/upholding-professional-standards/sector-standards/valuation/red-book/ [Accessed 7 April 2020].

RICS, 2020o. Best Practice for Registered Valuers. [Online] Available at: https://www.rics.org/uk/upholding-professional-standards/regulation/valuer-registration/best-practice-for-registered-valuers/ [Accessed 8 Octrober 2020].

RICS, 2020p. Professional Indemnity Insurance Requirements, London: RICS.

RICS, 2020q. Decision Tree. [Online] Available at: https://www.rics.org/globalassets/rics-website/media/upholding-professional-standards/standards-of-conduct/ethics-decision-tree-rics.pdf [Accessed 2020 October 2020].

Index

[**NB** Page numbers in *italics* refer to figures, page numbers in **bold** refer to tables.]